Migration in Performance

This book follows the travels of *Nanay*, a testimonial theatre play developed from research with migrant domestic workers in Canada, as it was recreated and restaged in different places around the globe. This work examines how Canadian migration policy is embedded across and within histories of colonialism in the Philippines and settler colonialism in Canada. Translations between scholarship and performance – and between Canada and the Philippines – became more uneasy as the play travelled internationally, raising pressing questions of how decolonial collaborations might take shape in practice. This book examines the strengths and limits of existing framings of Filipina migration and offers rich ideas of how care – the care of children and elderly and each other – might be rethought in radically new ways within less violently unequal relations that span different colonial histories and complex triangulations of racialised migrants, settlers and Indigenous peoples.

This book is a journey towards a new way of doing and performing research and theory. It is part of a growing interdisciplinary exchange between the performing arts and social sciences and will appeal to researchers and students within human geography and performance studies, and those working on migration, colonialisms, documentary theatre and social reproduction.

Caleb Johnston is a Lecturer in Human Geography at Newcastle University.

Geraldine Pratt is a Professor of Geography and Canada Research Chair in Transnationalism and Precarious Labour at the University of British Columbia.

Routledge Research in Place, Space and Politics
Series Editor: Professor Clive Barnett, University of Exeter, UK

This series offers a forum for original and innovative research that explores the changing geographies of political life. The series engages with a series of key debates about innovative political forms and addresses key concepts of political analysis such as scale, territory and public space. It brings into focus emerging interdisciplinary conversations about the spaces through which power is exercised, legitimised and contested. Titles within the series range from empirical investigations to theoretical engagements and authors comprise of scholars working in overlapping fields including political geography, political theory, development studies, political sociology, international relations and urban politics.

The Politics of Settler Colonial Spaces
Forging Indigenous Places in Intertwined Worlds
Edited by Nicole Gombay and Marcela Palomino-Schalscha

Direction and Socio-spatial Theory
A Political Economy of Oriented Practice
Matthew G. Hannah

Postsecular Geographies
Re-envisioning Politics, Subjectivity and Ethics
Paul Cloke, Christopher Baker, Callum Sutherland and Andrew Williams

Oil, Culture and the Petrostate
How Territory, Bureaucratic Power and Culture Coalesce in the Venezuelan Petrostate
Penélope Plaza Azuaje

Migration in Performance
Crossing the Colonial Present
Caleb Johnston and Geraldine Pratt

For more information about this series, please visit: www.routledge.com/series/PSP

Migration in Performance

Crossing the Colonial Present

Caleb Johnston and Geraldine Pratt

Routledge
Taylor & Francis Group
LONDON AND NEW YORK

First published 2019
by Routledge
2 Park Square, Milton Park, Abingdon, Oxon OX14 4RN

and by Routledge
52 Vanderbilt Avenue, New York, NY 10017

Routledge is an imprint of the Taylor & Francis Group, an informa business

British Library Cataloguing-in-Publication Data
A catalogue record for this book is available from the British Library

Library of Congress Cataloging-in-Publication Data
A catalog record has been requested for this book

ISBN: 978-1-138-88563-9 (hbk)
ISBN: 978-1-315-71533-9 (ebk)

Typeset in Times New Roman
by codeMantra

To migrant families, that their stories be heard

Contents

Figures

Acknowledgements

In a project that has unfolded over one-quarter century, we have accumulated many debts. We apologise in advance that this list is inevitably incomplete. None of this would have been possible without years of collaboration with the Philippine Women Centre (PWC) of British Columbia (BC). Although individuals within the PWC stand out for their incredible work and leadership over the years, we respect and honour their reluctance to be named individually and their commitment to the PWC as a collective. We thank the organisation as a whole. A number of the individuals who we first worked with at the PWC subsequently formed Migrante BC and we also thank the latter organisation for their support, for staging a conversation for the video component of the production at PETA Theater and for introducing us to their colleagues at Migrante International in Manila. Our heartfelt thanks to Migrante International for committing to our collaborative project and seeing it through to completion, against all odds. At a very busy time for their organisation, the Canadian Filipino Association of the Yukon (CFAY) generously extended their help. All of these Filipino migrant organisations are over committed, in Canada relying wholly on volunteer labour; we are indebted for their time, energy and acute theorising. We hope that the plays – in some measure – have been useful to them.

Each location and each production have their own set of debts. *Nanay's* travels in the world were only possible through many connecting threads, people, encounters and contributions. The productions would not have come into being without the help of many different people, and each production emerged through existing relationships and created new ones. The play undoubtedly would not have been created without Alex Lazaridis Ferguson, the director for the first three productions: in Vancouver, Berlin and the PETA Theater Center. We thank Alex for his creative direction, open collaboration, keen appreciation of space and enthusiasm for experimentation. Martin Kinch and others at the Vancouver Playwright Theatre Centre offered their much-needed wisdom and skill as dramaturgs. Thanks to Norman Armour, in his role as executive director of the PuSh International Performance Festival, for supporting and launching *Nanay*, and to the whole festival team. The PWC and FCYA (Filipino-Canadian

Youth Association) were integrally involved in the Vancouver production at all stages: in script development, creating the basement bedroom scene and the audio for the sound installation, working as interns with the stage manager and director, as guides within the performances, and coordinating the Filipino audiences. We thank Charlene Sayo and Carlos Sayo in particular for their participation throughout, including co-facilitating the talkbacks at the end of each performance. We thank Dinah Estigoy for travelling with us to take on this role in the Berlin production. In Berlin, we are grateful to (and still a little amazed by) the Hebbel am Ufer Theatre for finding *Nanay* on the Internet and including it in their Your Nanny Hates You! festival on the family. Special thanks to Stefanie Wenner and Anna Mülter for finding us, for their hospitality and for coordinating the production at the HAU Theatre. For the Vancouver production, we thank Meissa Dioniso, Alexa Divine, Lissa Neptuno, Patrick Keating, Karen Rae, and Hazel Venzon (the performers in the Vancouver production) and Andreas Kahre (scenographer), Tamara Roe (design), Noa Anatot (stage manager), Barbara Clayton (costume) and John Webber (lighting). In Berlin, we thank Delia Brett, Meissa Dioniso, Lissa Neptuno, Patrick Keating, Elia Kirby, Hazel Venzon and Erin Wells. Hazel deserves special thanks as she has moved with and beyond *Nanay*, generously bringing us along on some of her travels, in new and challenging directions.

We ventured to Manila in almost total ignorance. Thank you to Dennis Gupa for an early sketch of the theatre scene in Manila and preliminary ideas about possible presenters. Thanks to Dada Docot for an introduction to the PETA Theater Center and to Nora Angeles for introductions to a number of colleagues and friends in Manila. And although we eventually chose to present at the PETA Theater, our heartfelt thanks to Carolyn Sobritchea, then Dean of the Asian Center at the University of the Philippines, Diliman, for her generous welcome and continuing friendship and assistance over the years. We had the very good fortune to meet Vanessa Banta, then on faculty in the department of Speech Communication and Theater Arts at UP Diliman, on our first trip to the Manila. Vanessa attended rehearsals of the play and opened her graduate dramaturgy course for feedback and comment on its performance. She was then absolutely central to the Bagong Barrio production. We were immensely fortunate for the opportunity to work with PETA colleagues, especially programme director Queng Reyes and artistic director Isabel Legarda, and we are grateful to everyone who worked on the production: Marichu Belarmino, Alex Ferguson, Anj Heruela, Patrick Keating, Joanna Lerio, Lex Marcos, Hazel Venzon (actors); and David Kerr (stage manager and technical director). We thank May Farrales, a PhD candidate in geography at UBC at the time, and former member of the PWC in Vancouver, for arranging for family members to attend the play and for coordinating and conducting the interviews with family members. May also contacted and coordinated the attendance of migrant and other nonprofit organisations as well as government representatives. We are grateful to May and Teilhard Paradela, also a PhD candidate at UBC at the time and a Manila native, for co-facilitating the talkbacks after every PETA performance. Thanks to Jessica Hallenbeck for creating the video of the Migrante BC informal conversation, which was used in the PETA performance.

It should come as no surprise that gifted migrant organisers are also gifted collaborators, and Rina Anastacio, then of Migrante International, took the initiative that put the production in Bagong Barrio into motion and supported it (and us) through challenges. Allan Bonifacio, an exceptional organiser from whom we truly learned what it means to dedicate one's life to struggle, took responsibility for the production and for coordinating our research and time in Bagong Barrio. We thank Rommel Linatoc for directing the play and for working with us and the newly formed Sining Bulosan. The youth who committed to work on the production did so with dedication that was truly inspiring. A number did not make it to the end of the four-month process: one was called by his labour recruiter to return to Jeddah, for another his parent insisted that he resume working for wages. On the other hand, one member is now an active member of Migrante and is continuing the cultural work of the play. We are so grateful for our time with them, for their generosity of spirit and for what they taught us. We are grateful as well to Vanessa Banta, who was central to the entire production: translating the script to Tagalog and working alongside – and often far ahead of – us as a gifted organiser, interpreter, researcher, translator and friend. She co-authored the published material on which the Bagong Barrio chapter is based and we are grateful for her permission to rework the material here.

Tlingipino Bingo emerged out of our long-standing relationship with Hazel Venzon. By 2014, she had moved to Whitehorse and through her association with Nakai Theatre, coordinated a script reading as part of the Pivot Theatre festival in January 2015 in collaboration with the Yukon Arts Centre. (Read by Grace Estrella, Marivic Perlawan, Claire Ness, Suki Wellman, George Maratos and Fatima Bruckman, with community forum facilitated by Hazel and Yvonne Clarke.) Hazel invited us into her connections with CFAY and coordinated and collaborated on the interviews to create a new monologue for the play. Following, and as a result of conversations in the forum, Hazel and Sharon Shorty spent time in a short artist residency in Juneau in December to January 2015–2016, from which emerged the seeds of the idea for *Tlingipino Bingo*. We thank all members of the "Conversation Collective" (which included Dennis Gupa, Ardie Carbardo, Sharon Shorty, Ricky Tagaban, Hazel Venzon and ourselves) for their willingness to collaborate on this project. Thanks to Dennis Gupa for directing the performance, and to everyone at the Elks Lodge. Marivic Perlawan helped on a number of occasions: on the first visit in December 2014, when she both told her story and performed in the script reading, for her participation in *Tlingipino Bingo* and permission to use her image in this book. CFAY, as we have indicated, needs special thanks for their generosity and enthusiasm, for their hard work getting the Filipino community out, for catering for the event and for their delicious BBQ skewers. We thank Vanessa Banta and Kelsey Johnson for their research assistance.

We have the extraordinary privilege of living and working among gifted and generous friends and colleagues who have been willing to read parts of the manuscript, often on very short notice. These include Rina Anastacio, Trevor Barnes, Alistair Bonnet, Jill Bruetsh, Rachel Brydolf-Horwitz, Emilie Cameron, J.P. Catungal, Kate Coddington, Huey Shy Chau, Julie Cruikshank, Dia Da Costa, Michelle Daigle, Jessica Dempsey, Dennis Gupa, Jennifer Hyndman, Cindi Katz, Diyah Larasati, Jacquelyn Micieli-Voutsinas, Alison Mountz, Natalie Oswin, Rachel Pain, Raksha Pande, Katarina Pelzelmayer, Karin Schwiter and Juanita Sundberg, along with anonymous referees of previously published work. Scholarly work is a collective enterprise and individual chapters also have benefited from feedback from the UBC Economic Geography reading group, from participants at the 2009 Antipode Summer Institute at the University of Manchester, the Reimaging Creative Economy Workshop in Edmonton Alberta in May 2017, and from those who attended the reading of the play at the 2012 annual meetings of the Royal Geographical Society (with the IBG) in Edinburgh (read by Alisa Anderson, Kristin Murray, Julie Taudevin and Corey Turner). Thanks to Chris Philo for that invitation. We have also benefitted from a number of presentations in the UBC Philippine Studies Series, from presentations at geography departments at the University of Zurich, the University of Glasgow, University of Leeds, Exeter University and the Canadian Studies Program at Harvard University, and in response to the *Society and Space* lecture at the annual meetings of the American Association of Geographers in 2017. Thank you to Rachel Brydolf-Horwitz for help with copyediting.

The weight of our argument throughout the book has been that ideas have a materiality that moves in the world, which implies that material resources are also necessary. For the Vancouver production, we thank the generous support of the Social Sciences and Humanities Research Council (SSHRC), Canada Council for the Arts, Vancouver Foundation, City of Vancouver, British Columbia Arts Council and the PuSh Festival. For the PETA and Bagong Barrio productions, we thank the PETA Theater Center, SSHRC and UK Economic and Social Research Council. For *Tlingipino Bingo*, we thank UBC Faculty of Arts, SSHRC, the Yukon Arts Fund, Whitehorse Nuit Blanche and Culture Quest. We thank Vanessa Banta, Anna Crawford, David Oro and Hazel Venzon for permission to use their photographs.

An earlier version of Chapter 1 appeared as "Creating New Spaces of Politics: *Nanay: A Testimonial Play*" in Geraldine Pratt, *Families Apart: Migrant Mothers and the Conflicts of Labor and Love* (Minneapolis: University of Minnesota Press, 2012). Portions of Chapter 3 appeared as Geraldine Pratt, Caleb Johnston and Vanessa Banta, "A Travelling Script: Labor Migration, Precarity and Performance," *TDR The Drama Review* 61, no. 2 (2017): 48–70 and in Geraldine Pratt, Caleb Johnston and Vanessa Banta, "Life-times of

Disposability and Surplus Entrepreneurs in Bagong Barrio, Manila," *Antipode* 49, no. 1 (2017): 169–192. Chapter 4 is revised from Caleb Johnston and Geraldine Pratt, "*Tlingipino Bingo*, Settler Colonialism and Other Futures," *Environment and Planning D: Society and Space* 35, no. 6 (2017): 971–993. We are thankful for permission to rework this material here.

Thanks to Deb for her thoughtful questioning, to Callen and his worldly wonder, and to Kaya, a love often too far away. As ever, thanks to Tohmm, for endless listening and support.

Abbreviations

CBSA	Canadian Border Services Agency
CFAY	Canadian Filipino Association of Yukon
CFO	Commission of Filipinos Overseas
CIC	Citizenship and Immigration Canada
DFA	Department of Foreign Affairs
DOLE	Department of Labor and Employment
ECPAT	End Child Prostitution Child Pornography and Trafficking of Children for Sexual Purposes
FCYA	Filipino-Canadian Youth Association
HAU	Hebbel am Ufer
KPT	Kilusan para sa Pambansang Demokrasya
LCP	Live-In Caregiver Program
NGO	non-governmental organisation
OECD	Organisation for Economic Co-operation and Development
OFW	overseas Filipino worker
OUMWA	Office of the Undersecretary of Migrant Workers' Affairs
OWWA	Overseas Workers Welfare Administration
PETA	Philippine Educational Theater Association
POEA	Philippine Overseas Employment Administration
PWC	Philippine Women Centre
TFW	temporary foreign worker
UBC	University of British Columbia
UP	University of the Philippines
WTO	World Trade Organization
YNP	Yukon Nominee Program

Introduction

Labour migration and a travelling play

How can we understand a mother leaving her children with her sister in the Philippines to care for another family's children in Canada? Is it a question of survival? Evidence of the structured violence of racial capitalism, played out across the globe? A case of new digitally mediated global intimacies between parent and child? Of forced migration? Is it yet another repetition of colonial relations? Or is it evidence of personal ambition, freedom and fortitude, of finding one's luck? Given growing ethno-nationalism and opposition to migration in many countries, how we understand and tell such stories are of critical – even of life and death – importance.

But how to tell such stories? Who narrates? To whom? And to what effect? This book is an account of our attempts to move academic scholarship on labour migration into the world through public performance, to put what we understand to be the structural violence of state-sponsored forced migration more fully into public view. Collaborating with Filipino migrant organisations in Canada and the Philippines, we assembled audiences in shared performative spaces so that the issues could be experienced and felt, debated and addressed. Constructed from research transcripts, our play – entitled *Nanay* – has travelled beyond Canada to Berlin and Manila. This book follows its travels, to report on what we have learned and were taught along the way.

These journeys up-ended what we thought we were doing when we first created the play in Vancouver. We thought we were disseminating scholarly research on the lives and circumstances of Filipino migrant domestic workers in Canada to a public beyond the academy, to prompt more nuanced intercultural debate. The theatre professionals among us thought we were experimenting within the genre of testimonial theatre. What emerged was a new way of doing research and theory and a slow awkward process of unlearning or relearning about labour migration, our whiteness and racial formations within both colonialism and settler colonialism. The script of *Nanay* was already a complex translation of spoken words and gestures into paper (from interviews to transcripts into script), and as it has travelled to different times and spaces, it has invited new (re)iterations: new behaviours and meanings that have pushed against the limitations of our understanding of labour migration and drawn us to reimagine what scholarly knowledge is, can be and can do. In non-trivial ways, the play that we made has re-made us, in ways that we will explore. It has forced us to both recognise and reach beyond what has been termed methodological nationalism and to frame the migration of Filipino domestic workers within ongoing colonial relations in both the Philippines and Canada. It brought us to a different research practice, as part of a growing interdisciplinary exchange between the performing arts and social sciences, an exchange exploring possibilities situated at the intersections of research, performance and politics. These translations between scholarship and performance – and between Canada and the Philippines – became more and more uneasy as we journeyed alongside the play, and we hope that conveying some of our unease about narrative genre, authorship and authority can contribute to current thinking about what decolonial collaborations might mean in practice.

We put our play into the world with a pedagogical goal of bringing our research on the experiences of Filipino domestic workers in Canada to a wide transnational public. As *Nanay* travelled, we both shared and generated new and different stories and relations. We have returned home with a fuller understanding of both the strengths and limits of our framing of Filipina migration, and with ideas of how care – the care of children and elderly and each other – might be rethought in radically new ways within less violently unequal relations.

Origin stories

Nanay began at a crossroads of intersecting histories; and it was the merging of an interdisciplinary exchange of practices, networks and politics. One of us had been collaborating with the Philippine Women Centre (PWC) of British Columbia (BC) for over a decade to document the lives of Filipina migrant workers moving to Canada on temporary contracts as live-in domestic workers.[1] The Philippines now has one of the world's largest labour diasporas, with over 6,000 Filipinos leaving their families and country daily to work overseas.[2] Since 1992, women have outnumbered men among those leaving the Philippines as newly hired land-based legally deployed overseas Filipino workers (OFWs), the majority hired as domestic workers on temporary work contracts.[3] Although this has precipitated considerable debate within the Philippines about the effects of especially mothers' migration on the fate of their children,[4] there has been considerably less public discussion at the other end of what has been termed the global care chain. Often unfounded assumptions are made about the capacity of extended families in non-Western societies to absorb a migrant mother's absence,[5] and the timing and spacing of racialised labour migration keep the circumstances that drive migration and their intimate consequences for migrant families off side and out of view. In Canada, complacency around mothers leaving their children at home in the Philippines to care for others in Canada is also tied to the temporary worker programme through which these migrants arrive.

Canada's Live-in Caregiver Program (LCP) has been thought to be one of the most generous programmes of its type in the world because those registered in the LCP have had the opportunity to transition into permanent resident status after completing 24 months as a (registered) live-in caregiver (if able to complete this within a 48-month period). By 2007, almost 14,000 Filipinos (mostly women) were admitted annually to Canada through this programme, and the LCP had become a major point of entry for many Filipinos migrating to Canada.[6]

Research done in collaboration with the PWC added to what is now a large body of scholarship that documents conditions under the LCP and the ways that the programme lives on to marginalise Filipino women, their children and the Filipino community in Canada, principally through family separation and the deskilling of both adults and children. Though now a familiar scholarly story, typically told as the dehumanisation and exploitation of women from the global south and off-loading of the costs of social reproduction from global north to global south within a worldwide system of racial capitalism (that is, a system in which racial distinction is a structuring analytic and source of value),[7] it has limited reach and impact beyond the academy. This is one point of origin for our collaboration creating the testimonial play.

The other of us was rooted in the Vancouver theatre community and had been running a performing arts company with a history of documentary-based work. The interest and desire to work more fully within the genre of testimonial theatre was – in part – spurred by a serendipitous encounter with Nicholas Kent in 2006, then the director of London's Tricycle Theatre, who had been invited to Vancouver to speak about his work on *Guantanamo*, a 2004 documentary play staging the testimony and experiences of the family members of British citizens detained in the U.S. military prison in Guantanamo Bay, Cuba. A long-time colleague, Vancouver director Alex Ferguson, was drawn into the mix, and collectively we gleaned the importance and opportunity of working with a substantive archive of interviews collected in the research described above. We were drawn to the possibilities offered by testimonial theatre to place the issue of labour migration more fully in the public eye and were keen to draw on the capacity of live performance to forge and build engaged publics on the issues. To this end, we began working together with the PWC in 2007 to write a testimonial script based on verbatim monologues developed from existing research interview transcripts. The script would itself undergo several iterations, revisions and two workshops with professional actors before its public production at Vancouver's 2009 PuSh International Performing Arts Festival. We describe the process of creating the play more fully in the next chapter.

Situating *Nanay* in documentary performance

As we began translating interview materials and research archive into performance, we became aware that we were a small part of a growing international trend and focus. As Julie Salverson noted at that time, "Theatre that engages people's [actual life] stories is very fashionable these days,"[8] and there is now a vast literature on research-based theatre as a method of investigation that straddles theatre and social science research,[9] including works that focus on the lives of Filipino domestic workers.[10] By 2008, Tom Burvill was documenting a profusion of "performance responses" around the theme of migration and refugees, and no less than two special issues of *Research in Drama Education* (2008, 2018) have been dedicated to the topic of performance and refugees, along with a number of books, edited volumes and review articles addressing performance and international migration.[11] Alongside this scholarly attention, there has been a profusion of documentary-based artistic works examining the interconnections and intersections of migration, war, detention, state power and crisis.[12]

A complete list of such work exceeds this introduction, though much has been the object of productive critique. Take, as one example, the strong response made by the Director of the UK-based Detention Action, Jerome Phelps, to Ai Weiwei's 2016 mimetic restaging[13] of the widely circulated photograph of Alan Kurdi, a three-year-old Syrian refugee who drowned trying to cross the Mediterranean from Turkey with his family on September 2, 2015. The photograph of Kurdi's lifeless body found at the water's edge sparked global outrage and condemnation of European responses to refugees arriving by and dying at sea. Weiwei's positioning himself as Alan Kurdi, however, was seen by Phelps and others as little more than opportunistic and exploitative. Phelps nonetheless urges for new ways of mapping Europe's refugee "crisis" and finds a place for artists within this project: "If the old hierarchical spatial configurations are no longer sustainable, or are only sustainable with the violence of walls and razor wire, then there is a role for art to set out alternative ways of mapping our predicament."[14] And indeed, artists have staged a plethora of responses, some of which clearly extend performance beyond a one-off affective event or site, bringing into focus the long-term political commitments and realisations that are possible through critical creative interventions. Consider the *Grandhotel Cosmopolis*, the "social sculpture" created in Germany in 2013, which involved the renovation of an abandoned home for the elderly and its transformation into a "hotel" inhabited by refugees (hotel guests without asylum), visitors and artists, who live and work together, co-create artistic work and co-navigate state bureaucracies.[15] For Anika Marschall, the *Grandhotel Cosmopolis* helps us to rethink "performance's political efficacy" and demonstrates its movement "towards the real, towards social change."[16] Our work thus sits within an expansive body of critical artistic inquiry on migration in a time wherein migration has become a deeply acrimonious, divisive and politicised issue in the West and beyond.

Our play and process also sit within a tremendous interdisciplinary scholarly interest in performance and performance studies that dates from the 1990s,[17] with calls across the social sciences, the arts and humanities for radically "new" and experiential approaches to understanding social, economic and political life. Social scientists across a range of disciplines have been experimenting with a variety of creative forms and practices: what Sallie Marston and Sarah de Leeuw refer to as "a wide range of doing."[18] Although the liminality of performance is often exaggerated[19] and the force of Judith Butler's theorising of performativity is that most performances are iterations of a norm, there has been a hopefulness that performance can offer openings to something new, something unexpected.[20] Indeed, we framed our original presentation of *Nanay* within this emotional register and pitched it on its capacity to open up unexpected possibilities and to bring diverse audiences into a closer emotional proximity to the issues, in new and more direct ways. Promoting the play on a local radio station before it premiered at Vancouver's PuSh Festival in February 2009, we were asked, "It's the middle of winter. It's going to be cold. It's going to be wet. It's going to be dreary. Don't people want to go to the theatre to have a good time? Why should people see this production?"[21] Listeners were alerted to the intensity of emotional engagement they would experience while attending our play (through phrases such as "compelling," "engaging," "stories that draw you in," "heart breaking"). We evidently were successful in achieving the desired effect because at the end of the PuSh festival the Children's Choice Awards, determined by 12 children from a Vancouver-area elementary school, declared our play the "Saddest" of the 17 plays presented at the festival that year.[22] As documentary-based performance, the objective of the project has consistently been to pull audiences into the issues to affect them through all the tools and devices available in performance: stagings, atmospheres, playing with spatial relations of proximity and distance, speech, bodily sensation, objects, movements and touching details of individuals' lives.

The turn to performance and experimental approaches speaks to the disillusionment with a current (global) political predicament, the rise of populist politics both near and far, and a felt urgency to both embody and enliven imaginations of what is possible (and indeed to maintain a memory of radical alternatives of the past) and to contribute to informed, nuanced public discourse. Turning to artistic practice to reinvigorate political imagination in deteriorating and precarious conditions, Lauren Berlant, for instance, has argued that we live at a time of impasse, no longer believing in or holding to a collective vision of the good life. The impasse is not only a moment of danger but also one of possibility, when the rules and norms of habitation are unstable. People need, she believes, to detach from the "life-destructive forms of the normative political world" and reattach to public life and to politics in newly life-sustaining ways.[23] Inventing new genres and practices for doing research and theory is one part of the urgent process of refashioning publicness and idioms of political attachment. These are sites in which "movement happens through persons" often as a "space of abeyance … a soft impasse."[24]

Performance potentially acts as this kind of space of abeyance, as a temporary suspension from everyday life, a soft impasse for radical re-imagining and a time-space in which new social relations can come into being. It has, as Claire Bishop says of participatory art, a "double ontological status: it is both an event in the world, and at one remove from it."[25] Certainly not unique among the arts, theatrical performance also has the capacity to hold different times and places together in the same space, to enact a kind of heterotemporality. It is a space in which the present can be haunted by other times and places, and multiple ontologies, narratives, world views can coexist. These "crowded histories" can disrupt the singular historical narratives that so often foreclose the experiences and claims of the disadvantaged or dispossessed.[26] The ludic and embodied quality of live performance also holds the potential to prompt audiences to (re)think and engage their worlds differently and to extend the terms of political discussion in productive ways. This perhaps has been most fully articulated by Jacques Rancière who has developed many of his ideas about politics, democracy, egalitarianism and equality through theatrical metaphor,[27] so much so that Peter Hallward applies to his thinking the Platonic term, "theatrocracy."[28] As discussed at length in Chapter 1, Rancière is alert to theatre, not just as metaphor, but as a concrete space and practice, and as a privileged space of politics that can "redistribute the sensible." Not only the performance itself, but the relationships between actors and audience and between audience members themselves can scramble existing identifications and redistribute what we see, hear, think and say. Theatre holds radical possibility because it allows for – even encourages and plays with – disidentification and dissensus, and for Rancière, it is these moments that are full of democratic potential if taken-for-granted hierarchies of who can speak or be heard are momentarily suspended. They are the site wherein new political attachments are possible, where new identifications can be forged and possibly carried back out into the world. Live performance offers a powerful means and aesthetic space in which to play with proximity and distance, identification and disidentification, and to hold multiple – often conflicting perspectives – open and fluid.[29] And finally, less visible to the audience, but important nonetheless, performances are always – literally – physical coordinations taking place in time and space. *Nanay*, for instance, brought together ensembles of writers and researchers, activists and set designers, directors and actors, migrants and stage managers, technicians and audiences. "Whether cast in aesthetic or social terms," Shannon Jackson writes, "freedom and expression are not opposed to obligation and care, but in fact depend on each other; this is the daily lesson of any theatrical ensemble."[30] In our creative and scholarly work that places domestic work – freedom, unfreedom, obligation and care – at its core, this is no small consideration.

Scholars committed to the potentials of performance nonetheless hesitate over excessively celebratory approaches to the political and affective potential of performance, and we frame our project within this hesitation as well. Lauren Berlant has cautioned that the "affective turn emerges within the long neoliberal moment of the attrition of the social," and deep emotional responses, affective transformations and momentary connections can be a replacement for rather than a step toward sustained social engagement.[31] Artistic practice can be (and is) scripted into government or state projects that are anything but emancipatory.[32] With reference to the deployment of creative economy discourse in India, Dia Da Costa writes, "Unabashedly, this amounts to telling the hungry poor, who have no food or jobs or capital, that they could instead become creative entrepreneurs. Eat heritage (or, cake) if you don't have bread."[33] In our case, the first production of *Nanay* was indirectly funded by Cultural Olympiad monies provided by the City of Vancouver. Some local artists refused this funding because of what they perceived to be its effects on Vancouver's cultural scene: "a real trend of neutrality," "a switch [...] toward a market based creative atmosphere" and "a gutting of the artist run community."[34] And certainly, there is nothing inherently radical about conceiving theatre as an intensified site of circulating affect, intersubjective encounter and ethical communication. Quite the opposite, theatre as a heightened space of emotionality potentially maintains the hegemonic (and gendered) distinction between emotionally laden leisure time and artistic practice, and the rationalities of economic and political life. The problem is compounded when it is women from the global south who are the object of empathy in the global north. As we move through the book, we engage and re-engage our unease with the (geo)politics of using theatre to forge empathetic identification through the circulation of Filipina testimonies of hope, fear, trauma and separation and we reach towards more complex and compromised modes of empathy and engagement.

We are nonetheless alert to what theatrical performance does (or can do) throughout the following pages, with the proviso, in line with Claire Bishop, that much of the important work that performance does lies beyond what can be assessed through positivist sociological approaches and/or cultural policy think-tank and research assessment exercises that prioritise demonstrable "outcomes" or "impacts."[35] We will think as well with Shannon Jackson's observation about the work that is now emerging at the interface of the visual arts, performance and social engagement:[36] what is innovative depends on what one understands oneself to be disrupting. That is, it is innovative or disruptive from a particular disciplinary or vantage point. For social scientists working at the interface of research and performance, preconceptions about the intended or ideal audience of research might be productively disrupted. The process might unsettle assumptions about the desired outcomes of research, to the point that the need for – or ethics and politics of – translating performances back into text becomes an open question. Theatre professionals working at the interface of social science likely feel other kinds of innovating disruptions. Certainly, we felt the friction of working at the interface of theatre and social science throughout our process. In Vancouver, we experienced sometimes fraught creative disagreements about how to translate and perform a research archive. Discussions about authenticity and realism within testimonial or documentary theatre led us to rethink the fragility of social scientific claims to realism and objectivity.[37] The testimonial play both activated and disrupted concerns about statistical representativeness and generalisability. It forced us to take seriously whether and how academic research and writing provokes emotional responses in unintended and unacknowledged ways. It allowed us to hear the ways in which the contexts in which interview or testimonial materials are collected affect how stories are told: their tone, formality, phrasing and content. In Manila, the disruptions took shape in the same and different ways, often at the interface of postcolonial and neocolonial encounters and concerns. It was in Manila that we first realised that theatre could be a means of *doing* research otherwise, as a different mode of research practice. The testimonial monologues travelled uneasily into a new social, emotional and political terrain in the Philippines. What did it mean for us to move into, perform and to put stories of suffering experienced by migrants in Canada into conversation with family members left behind? What were the ethical implications as we tried to disrupt the fantasy of Canada as a dream destination? Innovating disruptions have varied as we have moved from research transcript to script, and from place to place, working at the interface of different questions and media. We trace these throughout the book.

Travelling alongside *Nanay*: an ethnography of worldly encounters

Amid the frenzy of the public production of *Nanay* at Vancouver's 2009 PuSh International Performing Arts Festival, we received an email from Anna Mülter and Stefanie Wenner, curators at Berlin's Hebbel am Ufer (HAU) Theatre. They were curating a series of performance works along the theme of family for their Your Nanny Hates You! Festival, and they had found *Nanay* online via a Google web search. The play – as a thing itself – had entered into and was circulating in the digital world. After reviewing documentation of the project, arrangements were made, and eventually, with a company of six actors, director, stage manager and representative from the PWC, we presented *Nanay* in a series of performances at the HAU Theatre in June 2009. Berlin was the first stop from Vancouver in the play's travels over the next eight years, the beginning of an itinerary that took it (and us) to Manila (twice, for two very different productions) and to the Canadian north (again, on two occasions). We follow these travels.

To say that the play took on a life of its own that overwhelmed the categories that ordered our understanding of ourselves and our scholarly work is to gesture towards the liveliness of things. Mediated by digital technologies and documentation, the play caught the attention of those at the HAU Theatre and brought us along on its travels. As Robin Bernstein writes, "a thing can invite behaviors that its maker did and did not envision."[38] It can "script" behaviour. This line of thinking, now known as "new materialism," resonates deeply and widely with performance studies, which, according to

Rebecca Schneider, "has long rummaged at the border erected between objects and subjects."[39] As an historian, Bernstein is drawn to the notion of "scriptive things" because it blurs subject, object and agency. Rather than a rigid document that prescribes action, a script is a "set of invitations that necessarily remain open to resistance, interpretation, and improvisation." It is a substance "from which actors, directors, and designers build complex, variable performances that occupy real time and space."[40] These performances can set other actions, affects, bodies, emotions and meanings into motion. The reactions of the Philippine embassy audience members that we encountered at an early talkback for instance (see Chapter 1), provoked and energised us to apply for funding to take the play to Manila. The helpful critique of Migrante International during our first production at the Philippine Educational Theater Association (PETA) in Metro Manila nudged us towards a collaboration with the organisation and the project's performance in the migrant-sending community of Bagong Barrio (see Chapter 3). An invitation took and reworked the project in the Canadian North, twice (see Chapter 4). And so on.

In this, we have found our way towards what the anthropologist George Marcus has called a "new ecology of knowledge production."[41] For Marcus, the reflexive turn in his field of anthropology, which he dates from the 1980s, has been overtaken or overshadowed by what he terms a transitive or recursive turn. What he means by this is that knowledge is now developed collaboratively, in different registers and forms, to speak beyond the academy to other publics with which the researcher wants and needs to be in conversation. Such engagement is integral to the research process or fieldwork itself. "The key innovation in method," he writes, "is that reception is folded into ethnographic strategies of inquiry," a process that he believes is now more prevalent than the academy "has a vocabulary for or genres of access to."[42] Marcus himself has been and is very interested in working with artists, alert to the implications for knowledge production. Whereas (even the transitive) anthropologist, he argues, tends to be inclined to the academy and the production of books and articles legible within that sphere, the artist is more inclined to "seek an extended network or 'archipelago' of additional receptions and viewing of the work," to show the work in other venues and to learn from its reception. This impetus to restage or reproduce, he notes, is in line with the idea of multi-sited ethnography but also a little different. While in a multi-sited ethnography an object or a process is followed across multiple sites, a restaging of the "same" artistic intervention or script involves expanding audiences and exploring ideas in different contexts with different micro-publics.[43]

We have found the line between multi-sited ethnography and exploring ideas through restaging across performance/interventions to be exceedingly blurry. Fifteen years of collaborative research with the PWC of BC in Vancouver provided an archive of material and the community support and impetus to develop the play, *Nanay*. As it has travelled, *Nanay* has itself become a site of research and theorisation; a site for telling and gathering new stories and embodied experiences. We have folded many of these into renewed scripts, refocused on the local experiences of global labour migrants in often radically new venues. As a project placing sites and peoples into conversation, it has opened opportunities and the necessity for additional research and conceptualisation alongside it. Research and performance have become the medium and material for each other.

This mode of working is also a mode of theory-making. As the anthropologists, Dominic Boyer and Cymene Howe, note, theory "thrives on the mobilization of transparticular" insights but how it travels and how it should travel are open, disputed questions.[44] These are pressing questions for all researchers and especially those working in the global south, with Indigenous communities or non-hegemonic groups. Too often theory has travelled within Eurocentric frames of reference, displacing and violently absorbing within colonial logics of Othering and a singular narrative of capitalist development, worlds that exist beyond these framings.[45] To "provincialize" Europe, Dipesh Chakrabarty has written, involves exploring how thought may be "renewed from and for the margins," with the understanding that "of course margins are as plural and diverse as the centers."[46] The task is not merely to pluralise theory towards a more cosmopolitan theoretical practice but to cultivate a different ethos of theoretical exchange. Boyer and Howe imagine this as a lateral "portable analytics": "rather than an optic from the antipodes [...], portable analytics follows ethnographic life, inviting a lateral meshwork of thought respun to create agile links, not only across space and time but across different kinds of intellectual practice, academic and otherwise, including political tactics, artistic interventions, and other varieties of creative assemblages." Boyer and Howe remind us that the Greek term, *theoros*, described an ambassador sent by the polis to witness ritual events in another city and that his travels always began and ended in a particular place – his home. Theory is thus both learning from a distance and from other places, and "tethered to the familiar epistemic space of home."[47] There is in this movement the possibility of unlearning or relearning one's home and its place in the world. The travels of *Nanay* afforded the opportunity to connect and learn from different nodes in a global care chain; to move laterally across and between different social worlds and audiences; and to bring our learnings on the issues back into a Canadian context.

This involved more than moving theoretical insights from one place to another. Travelling and researching through the play was also about seeing from different vantages, beyond the blinders of "methodological nationalism." It involved theorising beyond the national context of Canada. Travelling and researching through the play has allowed us to build a fuller understanding of spatio-temporal interconnections between various locations along the global care chain, connections that move through families, displacement from land, the almost universal devaluation of care-work, colonialism and settler colonialism, neo-liberal forms of capitalism, and race and class hierarchy. The LCP is directly parasitic, for instance, on the experience of American colonialism in the Philippines, as we will explore in Chapters 2 and 3. Researching through the play was about understanding the mutual processes of constituting migrant labour in Canada, the Philippines and other migration destinations, including the Middle East. The travels of *Nanay* have created the opportunity for a variant of what Gillian Hart calls relational comparison.[48] One does not need theatre, of course, to do this. But for us, theatre has worked well as a medium of exchange and passage across connected places and processes. Performance is in the world and part of the world differently and more fully than most academic texts and thereby puts us as researchers in the world more fully and differently as well. Shih notes the verbiness and activity in the method of relational comparison: "The entities brought together for comparison are, so to speak, *relationed*."[49] She follows Hannah Arendt in seeing works of art (in her case literature) as worldly (as in the world, part of the world and making the world) and sees relational comparison as an ethical practice of bringing marginalised texts and places into relation with the rest of world: it means "bringing into relation terms that have traditionally been pushed apart from each other due to certain interests, such as the European exceptionalism that undergirds Eurocentricism."[50] Without wishing to claim too much for *Nanay*, as it travelled it drew us into this active process and practice of relational comparison, of building up a more comprehensive sense of the relationships between temporary migration to Canada, labour export from the Philippines and the intertwined histories of migration and Indigenous dispossession in Canada. Much of the drama of and beyond *Nanay* emerged as and through relational comparison.

The project was never, however, just about returning home with a fuller understanding of labour migration and processes of racial capitalism. As documentary theatre, *Nanay* moves differently than an academic text, connecting people and narrative. Theatre is often visceral. It can punch you in the gut, send chills through your body, make you cry. We drew on and mobilised the capacity and intensity of performance to produce spaces of affective encounter, spaces to bring people together and invite into risky public conversations. As performance, the project is always about trying to find, embody, and perform translations and transnational connections. Consider the restaging of a monologue in which a young Filipina woman speaks about her troubled reunification with her mother in Vancouver. In the first production in Manila (see Chapter 2), the scene was performed by Manila-based actor – Anj Herula, who had no direct experience with the issue, insofar as none of her own family members have migrated overseas. Her relationship to the character and script developed very slowly. In rehearsal, she interpreted its emotional peak or crisis to be when she speaks very passionately about fighting with her mother in Canada and articulates her keen desire to return home to the Philippines. In the first actual performance, however, the actor was overtaken by a different "breaking point" in the monologue. Anj later recalled, "I got to the part where I was talking about 'we had a small goal to become a restaurant manager'" and broke into tears. "It was so strange for me, because it was an alien emotion compared to what I had rehearsed. I was actually kind of afraid of it because I didn't know if I would be able to handle where it would take me." The emotional gravity of the monologue thus radically and unexpectedly shifted: from fighting with her mother about wanting to return home to the Philippines (a response more easily understandable from her vantage in Manila) to inhabiting the character's profound alienation in Vancouver: the slow violent shrinking of aspirations from high performing university student in the Philippines to a dead-end job as a manager in a fast-food restaurant in Vancouver. Anj's experience speaks both to the existence of radical breaks in perspective and the possibility of perspectival shifts within and through performance. The problems associated with inhabiting another's story through empathetic engagement are a focus of attention throughout the book, in particular Chapter 1. But in this case, Anj's job as an actor was to inhabit a character and her perspective shifted, from the Philippines to Canada, to imagine the pain of diminished life chances. Such shifts and connections lay at the heart of our project.

Despite our embrace of "the recursive turn," there are reasons to pause over existing accounts of collaborations between researchers and artists, and portable analytics for theory building. Marcus describes the spaces that permit middle-range,[51] situated theory-making in fieldwork as para-sites; Michel Callon and Vololona Rabeharisoa name the hybridisation of knowledge and information exchange across scientific and other networks as "research in the wild."[52] While arresting and amusing, these terms and phrasings are also troubling, calling up and invoking the imagery of disease and colonial classification. And although Boyer and Howe's notion of transporting theory from one place to another through lateral networks of connection, in part to understand the familiar in new ways, bears some relation to the notion of *theoros*, an important difference is that the *theorio* were sent in response to an invitation by the city-state at which the ritual events were to be witnessed.[53] The issue of who has the capacity to travel and transport witnessed experience or knowledge – with or without an invitation – is power laden. Sara Ahmed roots her phenomenology of whiteness in an orientation "from which the world unfolds" before a body that trails behind unmarked and unstressed. Whiteness allows some bodies to move through space with ease and to "inhabit the world as if it were a home." "The politics of mobility," she writes, "who gets to move with ease across the lines that divide spaces, can be re-described as the politics of who gets to be at home, who gets to inhabit spaces, as spaces that are inhabitable for some bodies and not others, insofar as they extend the surfaces of some bodies and not others."[54] And there is, of course, a political economy that supports this differential travel. The production of *Nanay* in Vancouver was supported by the funding tied to the Vancouver Olympics. We could travel to Berlin and the Philippines and the Canadian North because of the availability of German and Yukon government cultural funding and our privileged access to the resources of our respective national funding agencies. The relative ease with which we moved across borders, translating and transporting narratives, and the mobility and privileges afforded to us by our whiteness are issues that kept reappearing, and were renegotiated and rethought through the travels of *Nanay*. Its performances were always and necessarily a site of invitation and appropriation, of shared anxieties and promises that prompted a series of uneasy collaborations and productive refusals to which we attend as this book unfolds.

Following the inclination of researchers noted by George Marcus, we have come full circle to produce this book: representing, translating, analysing, documenting, archiving embodied performance and social and economic life as academic text. This is, Diana Taylor has argued, a different way of knowing and being in the world than is a performative, embodied way of knowing.[55] The former she characterises as archive, the latter as repertoire. Repertoire is a form of embodied cognition, collective thinking, a knowing in place that requires presence, while archives are typically places and modes of knowing aligned with the sedimentation of power and maintenance of public order, classification, expertise, and linear and individualised thought. Archived memory and written texts can work across and at a distance, separating knowledge from its source. Reflecting on the impulse to write about the first performance (text that appears in revised form as the next chapter), the urge came in part as an effort to regain a kind of mastery over interpretation in an often-contested collective process of creation.[56] But as Taylor also notes, the archive and the repertoire are "not static binaries" and their relationship is "not by definition antagonistic or oppositional." Indeed, she is at pains to disrupt such a dichotomy, which could easily support a historicist and deeply Eurocentric narrative of tradition and modernity. The two forms of knowledge "exceed the limitations of the other" and "usually work in tandem" and "alongside other systems of transmission."[57] *Nanay* has been a site of multiple forms of knowledge transmission. While designed to perform an archive, in its movements, *Nanay* has picked up and generated new stories, effectively assembling a new archive that both has been folded into new iterations of the script and become the subject of academic inquiry and production. The lines between repertoire and archive have blurred. Throughout the book, we consider how we might write (as academics) in tandem and alongside public presentations of the play so as to sustain the active and disruptive potential of embodied performance. We are suggesting (or hoping) that there are moments when archiving performance can destabilise rather than sediment hegemonic power relations. We turn to introduce the collaborations through which this book has been built, as well as a thematic itinerary.

Uneasy collaborations and an itinerary

The research and processes we describe have been carried out in collaboration with Filipino advocacy organisations and activists. We draw as well on Shannon Jackson's sense that all performance-making is necessarily collaborative: "[T]o work in performance is to remember, and then forget and to remember again, that such inconvenience is the price paid for being supported."[58] Rather than an inconvenience, the research on which this book is based could not have been done and, more importantly, would not have been done without this collaboration and support. The research partnership with the PWC began in 1995 at a moment when it had become politically and ethically impossible within feminist scholarship for a white middle-class U.S. born, Canadian researcher to collect and circulate the stories of Filipina migrant workers on her own.[59] The PWC set the terms of our relationship, fully aware of the politics of and conditions required for such a collaboration, which included co-constructing the methodology and theorising together through continued conversation and validation. Nonetheless, the challenges of disseminating domestic workers' stories without betraying the political interests of the PWC and appropriating their narratives for academic career advancement have been ever present. The creation and travels of *Nanay* – performing migrant stories – has only added complexity to these concerns, and we have been forced to learn and relearn and make efforts to unlearn our privilege in different ways within each new iteration of research and performance. Rather than a celebratory story of expanding and seamless solidarity, we write about our slow and awkward process of receiving the lessons opened to us. We keep alive the possibility that writing about this process is a final betrayal.

Nanay has taught us at least three general things about collaboration. First, the term shelters a welter of meanings and processes. In a feminist world, it usually implies an effort to distribute power relations, decision-making and resources more horizontally. In the theatrical world, it can mean different lateral ways of working and co-creating but – often – eventually creative roles are and must be distributed in a more hierarchical ordering with the director taking centre stage, maintaining final creative authority. While not unproblematic, working in theatre can offer important lessons in authorial release to the claims of others, a process we experienced repeatedly as the play travelled. Second, there is a political economy to collaborations that is rarely discussed but central to their success (or lack thereof). The first collaboration with the PWC was no doubt facilitated by the fact that the researcher accessed and transferred research monies to the organisation in advance of the research (thus relinquishing a measure of control). This is a model that we attempted to replicate whenever possible with *Nanay*, with varying success. And third, collaborations can take place at different velocities which shape their engagements and politics. Richa Nagar locates ethical "knowledge making and knowledge makers in an unending process of building and sustaining long-term relationships." These relations, she argues, "are marked by complex and fluid co-constitution [...] through modes of connecting and trusting, living and being, troubling and creating that are often ignored in dominant forms of scholarly inquiry, including in genres categorised as 'ethnographic' and 'activist' research."[60] We write of our efforts to enact this ideal *and* of the moments when we hesitated, stumbled, became less sure of ourselves and failed, with the understanding that we have learned as much (if not more) from our failures as our successes and that others might benefit from learning from them as well. The research collaboration on which *Nanay* is based was a very slow and gradual one that evolved from 1995 to 2012, with one research project following another as the PWC saw more issues to address within their community. There was time to get to know each other and to build trust. The first production of *Nanay* in Vancouver was also relatively slow, beginning within partially overlapping established relationships. A two-week development workshop in 2007 was followed by another such process in summer 2008, and then another development/rehearsal period of the same length in winter 2009. Given the distances and logistics involved, the collaborations in subsequent productions have been shorter.

It is not just the length of the later collaborations that has added complexity. Certainly, the politics of witnessing the lives of Filipina domestic workers in Canada has always been at issue. The Vancouver production was understood as an effort to bring together an intercultural audience into a single theatrical space to witness and discuss the implications of Canada's need and capacity to benefit from global patterns of uneven development and the poverty that forces migration from the Philippines. But inducing Canadians to empathise with the suffering figure of the Filipina domestic worker is a fraught enterprise. A hallmark, self-aggrandising feature of the liberal bourgeois subject is precisely the capacity for such identification,[61] and black and postcolonial feminists have for many years raised hard questions about the politics and geopolitics of who testifies to and who witnesses narratives of pain, arguing that witnessing the suffering of racialised women often reinscribes hierarchies of race and geopolitical privilege.[62] Our efforts to block the familiar pleasures of liberal humanitarian responses of pity and guilt, and to bring non-Filipino audience members into a more self-critical and compromised mode of witnessing is a focus of sustained discussion in Chapter 1.[63]

The politics of witnessing changed fundamentally when we reversed the journey of migrant domestic workers and moved the play into the socio-political and discursive terrain of the Philippines. We tell of our efforts to rewrite the script for the production at PETA in Chapter 2: to construct a more complex narrative of Filipina agency, to bring the circumstances of domestic workers in Canada to the relatives of domestic workers in the Philippines, and to put a view from Canada into conversation with one rooted in the Philippines. That this was only a partial success taught us lessons about colonialism, whiteness, and both our and the Filipino-Canadian diaspora's place within them. Our journey on this slow process of learning is evident, we hope, in our citational practices, which become somewhat less Eurocentric from this moment in the text. Who we cite and with whom we enter into conversation with within our scholarly texts is another face of collaboration.[64]

The next production of the play, described in Chapter 3, was (with the exception of one scene) translated into Tagalog. It was also substantially rewritten and revised by the Manila-based director, Rommel Linatoc. This productively reconfigured our relationships of power, dependency, equality and reciprocal exchange. Carol McGrahahan has observed that refusal is a concept that is "in dialogue with exchange and equality"; it can be "an element of group morality, a generative act, a rearrangement of relations rather than an ending of them."[65] The director's refusal to take our script as given and to insist on weaving into it his own script and theatrical practice reordered the terms and process of our exchange. The renewed play drew us into different theatrical traditions, aesthetics and other ways of grasping and representing the issues. In this reiteration of *Nanay*, the process of research and creation was relocated to a poor migrant-sending community in metro Manila and was refocused on the predicaments of migrant workers worldwide and the forces within the Philippines that reproduce them. It installed Canada's LCP as a minor player within a global context, as one of many migration destinations. Audience members were asked to witness, not only the precarity of migrant workers throughout the world, but the production of precarity within their own neighbourhood, through the subcontractualisation of jobs, the gutting of unions and the vagaries of state promises to land titling.

And finally, a Tlingit witness to a performed script reading in Whitehorse, in the Canadian North, opened us to a new collaboration, with a Filipino artist and two Tlingit performers, and grounded *Nanay* and an analysis of the LCP within another framing, ever present but previously invisible to us. This is the relation between the marginalisation of racialised migrant workers and Indigenous peoples in Canada, and the workings of racial capitalism within the frame and logics of settler colonialism. The Tlingit witness grounded the basis for a conversation in incommensurate but resonant experiences of the violence of state-imposed family separation. Like so many migration researchers and committed migration activists, our research and *Nanay* had been framed within discourses of citizenship and labour, erasing other claims and stories. As Leti Volpp writes, "Immigration is responsible for indigenous dispossession. But it also provides the alibi. [… I]t functions as both the reason for – and the basis of – denial."[66] In Chapter 4, we explore how the Filipino community in Whitehorse has functioned as a kind of alibi within settler colonialism in the Canadian North and introduce a performance that emerged in collaboration with Tlingit artists to explore resonances across different colonial experiences, opening the possibilities for imagining new, less destructive ways of co-inhabiting in Canada.

This book thus traces a trajectory from intercultural exchange in Canada, to thinking more deeply about the imbrication of Canadian immigration and care policies within the Philippine colonial experience, to reassessing racialised labour migration and settlement within ongoing processes of settler colonialism. This is the political-theoretical-personal journey that the thing, the script, set in motion. It is a journey that traces a path from Canada to the Philippines and back to Canada at a different place of return, one that productively unsettles our originating narrative of citizenship, exclusion and victimisation, and – more and more as we travel – our place as narrators of this story.

The 2017 film *Nervous Translation* tells the story of Filipino labour migration at a moment of transition from the authoritarian regime of Ferdinand Marcos to the presidency of Cory Aquino. The big national story of a massive programme of labour export is told from the perspective of a shy eight-year-old girl, Yael. The film spends time with Yael in her afternoons at home alone after school. Her unhappy mother works an exhausting job at a local shoe factory and her father has been working abroad in Riyadh for years. The young girl spends her afternoons listening to the tapes that her father has sent to her mother, and creating in miniature on her kitchen playset the scenes of domesticity longed for by her, her mother and her father. Yael attempts to order a world she cannot control and to make sense of events and emotional currents that she cannot understand. Translations run through the film. Yael carefully reenacts her father's words in her miniaturised domestic world. On a lined sheet of paper, she patiently translates the sounds of her home onto a partial list of English and Tagalog words: telephone: Rrrrgggg; ref[rigerator]: grrrrrrrr; air con: kash nugggg. Reading and taking literally an advertisement for a pen that offers the possibility of "a beautiful life," in her loneliness, she sets off to acquire this transformational tool for her translation work, only to find that she has insufficient funds. The filmmaker, Shireen Seno, gently indicates that parts of the story remain untold: we never know, for instance, why Yael has bandaging on her elbows.

Our book traces a series of uneasy translations, in the reverse, starting with the pen that promises so much and moving words back into life, translating from English to Tagalog, back to English. Scholars who order the world through possession of a magical pen rarely display the narrative restraint displayed in Seno's film and it is a lesson that we have learned slowly and trace throughout this book, ever curious about what is gained and lost through translation, and what can be learned at the interface of archive and performance, moving between different places in the world.

Notes

1 Geraldine Pratt, *Working Feminism* (Philadelphia: Temple University Press, 2004); Geraldine Pratt, *Families Apart: Migrant Mothers and the Conflicts of Labor and Love* (Minneapolis: University of Minnesota Press, 2012).

2 Migrante International, "#SONA2015 Number of OFWs Leaving Daily Rose from 2,500 in 2009 to 6,092 in 2015," *Migrante International* (blog), 29 July 2015.

3 According the UN International Labour Organisation, domestic workers are some of the most likely to face abuse and exploitation in their place of work, and the International Trade Union Confederation estimates that 2.4 million domestic workers, many of whom are Filipino, face conditions of slavery in the

Gulf Cooperation Council countries alone (see International Trade Union Confederation "Facilitating Exploitation: A review of Labour Laws for Migrant Domestic Workers in Gulf Cooperation Council Countries," Legal and Policy Brief, Brussels, 2014).

4 Rhacel Parreñas has written extensively about the "vilification" of mothers who leave their children to work abroad, not only in the Philippines, but in Poland where left children are termed "euro-orphans"; in Romania where mothers working abroad has been framed as a "national tragedy"; and in Sri Lanka. (Rhacel Salazar Parreñas, "Transnational Mothering: A Source of Gender Conflicts in the Family," *North Carolina Law Review* 88 (2010): 1825–1856.) As Joanna Dreby notes in relation to Mexican migrants, "Migrant mothers bear the moral burdens of transnational parenting." (Joanna Dreby, *Divided by Borders: Mexican Migrants and their Children* (Berkeley: University of California Press, 2010), 204.) Parreñas' claims about the vilification of migrant mothers in the Philippines are not uncontested. Filomeno Aguilar presents evidence from Batangas Province in the Philippines to demonstrate that "Unlike middle-class opinion makers based in Metro Manila," the people in the village in which he conducted his ethnography "refrain from passing judgment on transnational families and their growing children, especially adolescents, saying a lot depends on the individual child." (Filomeno Aguilar, "Brother's Keeper? Siblingship, Overseas Migration, and Centripetal Ethnography in a Philippine Village," *Ethnography* 14, no. 3 (2013): 352.) So too he argues that there is no stigma attached to being a child of migrant parents, "as may be the case elsewhere" (Ibid., 352; see also Deirdre McKay, *Global Filipinos: Migrants' Lives in the Virtual Village* (Bloomington: Indiana University Press, 2012).

5 Rhacel Salazar Parreñas, "Transnational Mothering: A Source of Gender Conflicts in the Family," *North Carolina Law Review* 88 (2010): 1825–1856.

6 The Philippines was by then the largest source country of immigrants to Canada. Between 2010 and 2014, 31,791 principal applicants and 28,872 dependents were admitted to Canada as permanent residents through the LCP, and since 2005 at least 25% (and in some years up to 45%) of all immigrants from the Philippines have come through the LCP (Citizenship and Immigration Canada 2014). The LCP was restructured as the Caregiver Program in 2014 and the numbers admitted have decreased substantially (e.g., just 3,983 caregivers were admitted nationwide on temporary work permits in 2015. See Government of Canada, "Temporary Foreign Worker Program Labour Market Impact Assessment (LMIA) Statistics Fourth Quarter, 2015," Quarterly Labour Market Impact Assessment Statistics, Ottawa, 2016). However, the transition from LCP to permanent resident status has become more precarious. Estimates in recent years are that of the more than 7,000 people with Caregiver visas each year only 500–600 have made the transition to permanent status annually (Saunders, 2018). The government introduced two strands to the programme in 2014, one for childcare and another dedicated to care of the elderly and those with chronic medical needs. The two streams have somewhat different requirements and there is a quota of 2,750 places for permanent resident status annually for each strand of the programme, suggesting that the programme no longer provides a clear pathway to citizenship: it offers the possibility but no guarantee to permanent residency. Time towards the required 24 months as a caregiver cannot be carried across the two strands of the programme, and if a caregiver is unable to complete the 24 months in a 48-month period, she cannot return to Canada for four years (a new "four years in, four years out" rule). Meanwhile, waiting times for processing permanent resident status have increased: as of January 2015, waiting times to gain permanent resident status *after* finishing

the programme had stretched to 50 months and at the beginning of 2018 there was a backlog of more than 30,000 Caregiver visa holders awaiting a decision on permanent residence. (See Nicholas Keung, "Foreign caregivers face lengthy wait for permanent resident status," *Toronto Star*, 21 July 2015, online, and Doug Saunders, "Family Ties," *Globe and Mail*, 16 June 2018. O1, O6–O7.) The issues of deskilling and family separation, therefore, remain and have likely worsened.

7 For example, Sedef Arat-Koç, "Politics of the Family and Politics of Immigration in the Subordination of Domestic Workers in Canada," in *Family Patterns, Gender Relations*, 2nd edition, ed. Bonnie Fox (Toronto: Oxford University Press, 1993), 352–374; Abigail Bakan and Daiva Stasiulis, *Not One of the Family: Domestic Workers in Canada* (Toronto: University of Toronto Press, 1997); Patricia Daenzer, *Regulating Class Privilege: Immigrant Servants in Canada, 1940s–1990s* (Toronto: Canadian Scholars' Press, 1993); Kim England and Bernadette Stiell, "'They Think You're as Stupid as Your English Is': Constructing Foreign Domestic Workers in Toronto," *Environment and Planning A* 29, no. 2 (1997): 195–215; Philip Kelly, Stella Park, Conely de Leon, and Jeff Priest, "Profile of Live-in Caregiver Immigrants to Canada, 1993–2009," TIEDI (Toronto Immigrant Employment Data Initiative) Analytical Report 18, Toronto, 2011; Jah-Hon Koo and Jill Hanley, "Migrant Live-in Caregivers: Control, Consensus, and Resistance in the Workplace and the Community," in *Unfree Labour? Struggles of Migrant and Immigrant Workers in Canada*, eds. Aziz Choudry and Adrian Smith (Oakland: PM Press, 2016); Audrey Macklin, "Foreign Domestic Worker: Surrogate Housewife or Mail Order Servant?" *McGill Law Journal* 37, no. 3 (1992): 681–760; Maria Wallis and Wenona Giles, "Defining the Issues on Our Terms: Gender, Race and State – Interviews with Racial Minority Women," *Resources for Feminist Research* 17, no. 3 (1988): 43; Geraldine Pratt, *Working Feminism* (Philadelphia: Temple University Press, 2004); Geraldine Pratt, *Families Apart: Migrant Mothers and the Conflicts of Labor and Love* (Minneapolis: University of Minnesota Press, 2012); Tanya Schecter, *Race, Class, Women and the State: The Case of Domestic Labour in Canada* (Montreal: Black Rose Books, 1998); Ethel Tungohan, *From the Politics of Everyday Resistance to the Politics from Below: Migrant Care Worker Activism in Canada* (Champaign: University of Illinois Press, forthcoming).

8 Julie Salverson, "Taking Liberties: A Theatre Class of Foolish Witnesses," *Research in Drama Education* 13, no. 2 (2008): 247.

9 For instance, George Belliveau and Graham Lea, eds., *Research-Based Theatre: An Artistic Methodology* (Bristol: Intellect, 2016); Johnny Saldaña, *Ethnotheatre: Research from Page to Stage* (New York: Routledge, 2011).

10 See Prue Wales, "Temporarily Yours: Foreign Domestic Workers in Singapore," in *Research-Based Theatre: An Artistic Methodology*, eds. George Belliveau and Graham Lea (Bristol: Intellect, 2016), 147–161. In Canada, another play on Filipino domestic workers in the LCP opened in 2010: *Future Folk*, created by Sulong Theatre Collective (see www.sulongtheatre.com/productions.html). After performing in *Nanay*, actor-director Hazel Venzon has created a number of performance works on the theme of Filipino migration to Canada: *Embrace* (2010) https://whatsupyukon.com/Yukon-Arts-Entertainment/arts/hazel-venzon-embraces-the-filipino-community/; *The Places We Go* (2018) http://unitprod.ca/the-places-we-go/.

11 See Tom Burvill, "'Politics Begins as Ethics': Levinasian Ethics and Australian Performance Concerning Refugees," *Research in Drama Education* 13, no. 2 (2008): 233–243; Emma Cox, *Staging Asylum: Contemporary Australian Plays about Refugees* (Redfern: Currency Press, 2013); Emma Cox, *Theatre & Migration* (New York: Palgrave Macmillan, 2014); Emine Fisek, *Aesthetic Citizenship:*

Immigration and Theater in Twenty-First-Century Paris (Evanston: Northwestern University Press, 2017); Rand Hazou, "Performing Manaaki and New Zealand Refugee Theatre," *Research in Drama Education* 23, no. 2 (2018): 228–241; John McCallum, "CMI (A Certain Maritime Incident): Introduction," *Australasian Drama Studies* 48 (2006): 136–142; Charlotte McIvor, *Migration and Performance in Contemporary Ireland: Towards a New Interculturalism* (London: Palgrave Macmillan, 2016); Stephen Elliot Wilmer, *Performing Statelessness in Europe* (London: Palgrave Macmillian, 2018).

12 In a comprehensive review, Emma Cox and Caroline Wake note that the locus of much artistic production has moved (over the past decade) from Australia to Europe in response to the geographies of ongoing humanitarian crises. (See Emma Cox and Caroline Wake, "Envisioning Asylum/Engendering Crisis: Or, Performance and Forced Migration Ten Years On," *Research in Drama Education* 23, no. 2 (2018): 137–147.)

13 See www.theguardian.com/artanddesign/2016/feb/01/ai-weiwei-poses-as-drowned-syrian-infant-refugee-in-haunting-photo.

14 Jerome Phelps, "Why is So Much Art about the 'Refugee Crisis' So Bad?" *openDemocracy 50.50*, 11 May 2017.

15 For more on this project, see https://grandhotel-cosmopolis.org/de/.

16 Anika Marschall, "What Can Theatre Do about the Refugee Crisis? Enacting Commitment and Navigating Complicity in Performative Interventions," *Research in Drama Education* 23, no. 2 (2018): 162.

17 See Dorinne Kondo, "Bad Girls: Theatre, Women of Colour and the Politics of Representation," in *Women Writing Culture*, eds. Ruth Behar and Deborah Gordon (Berkeley: University of California Press, 1995), 49–64; Dorinne Kondo, "The Narrative Production of 'Home', Community, and Political Identity in Asian American Theatre," in *Displacement, Diaspora and Geographies of Identity*, eds. Smadar Lavie and Ted Swendenburg (Durham: Duke University Press, 1996), 97–117; Dorinne Kondo, "(Re)visions of Race: Contemporary Race Theory and the Cultural Politics of Racial Crossover in Documentary Theatre," *Theatre Journal* 52 (2000): 81–107; Peggy Phelan, *Unmarked: The Politics of Performance* (New York: Routledge, 1993); Joseph Roach, *Cities of the Dead: Circum-Atlantic Performance* (New York: Columbia University Press, 1996).

18 See Sallie Marston and Sarah de Leeuw, "Creativity and Geography: Towards a Politicized Intervention," *Geographical Review* 103, no. 2 (2013): xxi. Within our own intellectual discipline of human geography, there continues to be sustained interest in the arts as a means for envisioning and enacting new ways of speaking, feeling and doing geographical knowledges, for example, see Leonora Angeles and Geraldine Pratt, "Empathy and Entangled Engagements: Critical-Creative Methodologies in Transnational Spaces," *GeoHumanities* 3, no. 2 (2017): 269–278; Sarah De Leeuw, *Where it Hurts* (Edmonton: NeWest Press, 2017); Harriet Hawkins, *For Creative Geographies: Geography, Visual Arts and the Making of Worlds* (New York: Routledge, 2014); Harriet Hawkins and Elizabeth Straughan, eds., *Geographical Aesthetics: Imagining Space, Staging Encounters* (London: Routledge, 2015); Sallie Marston and Sarah de Leeuw, "Creativity and Geography: Towards a Politicized Intervention," *Geographical Review* 103, no. 2 (2013): iii–xxvi; Mike Pearson, *Marking Time: Performance Archaeology and the City* (Exeter: University of Exeter Press, 2013); David Pinder, "Arts of Urban Exploration," *Cultural Geographies* 12 (2005): 383–411; David Pinder, "Sound, Memory and Interruption: Ghosts of London's M11 Link Road," in *Cities Interrupted: Visual Culture and Urban Space*, eds. Shirley Jordan and Christopher Lindner (London: Bloomsbury Academic, 2016), 65–83; Ruth Raynor, "Speaking, Feeling, Mattering: Theatre as Method and Model for Practice-Based, Collaborative, Research," *Progress in Human Geography*

early view (2018): n.p.; Amanda Rogers, "Advancing the Geographies of the Performing Arts: Intercultural Aesthetics, Migratory Mobility and Geopolitics," *Progress in Human Geography* 42, no. 4 (2017): 549–568. For a small sampling of those collaborating with or themselves producing creative works, see Tim Cresswell, "Displacements – Three Poems," *Geographical Review* 103, no. 2 (2013): 285–287; Paula Crutchlow and Helen V. Jamieson, "make-shift," *Liminalities: A Journal of Performance Studies* 10, no. 1 (2014): n.p; Sarah De Leeuw, *Geographies of a Lover* (Edmonton: NeWest Press, 2012); Sarah De Leeuw, *Where it Hurts* (Edmonton: NeWest Press, 2017); Neville Gabie, Joan Gabie, and Ian Cook, "Dust, in 'Bideford Black, the Next Generation,'" exhibition at the Burton Art Gallery & Museum, Bideford, 3 October 2015–13 November 2015; Michael Gallagher, "Sounding Ruins: Reflections on the Production of an 'audio drift'," *Cultural Geographies* 22, no. 3 (2015): 467–485; Richa Nagar in journey with Parakh Theatre and Sangtin Kisaan Mazdoor Sangathan, *Hungry Translations: Relearning the World Through Radical Vulnerability* (Champaign: University of Illinois Press, forthcoming); Cecile Sachs Olsen, "Collaborative Challenges: Negotiating the Complicities of Socially Engaged Art within an Era of Neoliberal Urbanism," *Environment and Planning D: Society and Space* 36, no. 2 (2018): 273–293; Michael Richardson, "Theatre as Safe Space? Performing Intergenerational Narratives with Men of Irish Descent," *Social & Cultural Geography* 16, no. 6 (2015): 615–633.

19 For instance, Jon McKenzie, "Gender Trouble: (The) Butler Did It," in *The Ends of Performance*, eds. Peggy Phelan and Jill Lane (New York: New York University Press, 1998), 217–235.

20 Enthusiasm about the potential of performance flows in part from the understanding that much of life is non-discursive and that the creativity of everyday life might be found in its embodied, performative aspects (Dwight Conquergood, "Rethinking Ethnography: Towards a Critical Cultural Politics," in *The SAGE Handbook of Performance Studies*, eds. D. Soyini Madison and Judith Hamera (London: SAGE, 2006), 351–365). Many social theorists have been drawn to affect theory because it traces both assemblages of norms, classifications and fixed hierarchical arrangements *and* affective intensities and atmospheres that are excessive to them, intimating new connections and encounters, and new ways of understanding and inhabiting the world. Considering the "intimate impacts of [affective] forces in circulation," anthropologist Kathleen Stewart writes, "They're not exactly 'personal' but they sure can pull the subject into places it didn't exactly 'intend' to go" (Kathleen Stewart, *Ordinary Affects* (Durham and London: Duke University Press, 2007), 21). They can produce "hard lines of connection and disconnection and lighter, momentary affinities and difference. Little worlds proliferate around everything and anything at all" (Ibid., 40).

21 Ironically, the radio programme, aired on Vancouver Cooperative Radio, is called Arts Rational.

22 For this, we received a women's rugby trophy dipped in chocolate, decorated with gold and blue stars. The Children's Choice Awards was itself a performance conceived and executed by the Mammalian Diving Reflex, a Toronto-based "research-art atelier." Twelve children from the municipality of Surrey's Bridgeview Elementary School were driven from event to event, and an award ceremony was held on the last afternoon of the festival. The children were given the option to present awards from more than 50 categories, only some of which touched on affective categories such as the most emotional, most joyful, most annoying, most horrific. Other categories included the longest tongue, biggest eyes, best costumes, best music. The children awarded in 25 categories overall: best beginning, best hair, most educational for parents, funniest voice, best use of technology, most "I don't care" (we're happy not to have received that), best

solo, most sad, most funny, most loud and clear, best use of adjectives, the longest, best teamwork, best use of props, creepiest, most interesting, most international languages, most weird, most realistic, most cheering applause, best driver, best overall (see http://childrenschoiceawards.blogspot.com).

23 Lauren Berlant, *Cruel Optimism* (Durham: Duke University Press, 2011): 229.

24 Ibid., 230.

25 Clare Bishop, *Artificial Hells: Participatory Art and the Politics of Spectatorship* (London: Verso, 2012), 274.

26 Erin Collins, "Of Crowded Histories and Urban Theory," Unpublished paper.

27 Jacques Rancière, *The Politics of Aesthetics: The Distribution of the Sensible*, trans. Gabriel Rockhill (London: Continuum, 2004); Jacques Rancière, *The Hatred of Democracy*, trans. Steve Corcoran (London: Verson, 2006).

28 Peter Hallward, "Staging Equality: On Rancière's Theatrocracy," *New Left Review* 37 (2006): 109–129.

29 See Erika Fischer-Lichte, *The Transformative Power of Performance: A New Aesthetics* (New York: Routledge, 2008); Clare Bishop, *Artificial Hells: Participatory Art and the Politics of Spectatorship* (London: Verso, 2012).

30 Shannon Jackson, *Social Works: Performing Art, Supporting Publics* (New York and London: Routledge, 2011), 14.

31 Lauren Berlant, "Thinking about Feeling Historical," *Emotion, Space and Society* 1, no. 1 (2008): 6.

32 See Dia Da Costa, *Politicizing Creative Economy: Activism and a Hunger Called Theater* (Urbana: University of Illinois, 2016); Laura Levin and Kim Solga, "Building Utopia: Performance and the Fantasy of Urban Renewal in Contemporary Toronto," *TDR: The Drama Review* 53, no. 3 (2009): 37–53.

33 Dia Da Costa, "Eating Heritage," paper presented at Reimagining Creative Economies workshop, Edmonton, Alberta, April 2017: 3.

34 Dawn Paley, "Olympics Cash and Vancouver's Cultural Community: Lines are being drawn between those who accepted Cultural Olympiad money, and those who refused it," *Vancouver Media Co-op*, 11 February 2010.

35 Clare Bishop, *Artificial Hells: Participatory Art and the Politics of Spectatorship* (London: Verso, 2012), 7.

36 Shannon Jackson, *Social Works: Performing Art, Supporting Publics* (New York and London: Routledge, 2011).

37 Geraldine Pratt and Caleb Johnston, "Translating Research into Theatre: *Nanay*: A Testimonial Play," *BC Studies* 163 (2009): 127.

38 Robin Bernstein, *Racial Innocence: Performing Childhood from Slavery to Civil Rights* (New York: New York University Press, 2011): 13.

39 Rebecca Schneider, "New Materialism and Performance Studies," *TDR: The Drama Review* 59, no. 4 (2015): 10.

40 Robin Bernstein, *Racial Innocence: Performing Childhood from Slavery to Civil Rights* (New York: New York University Press, 2011): 12.

41 George Marcus, "The Legacies of *Writing Culture* and the Near Future of the Ethnographic form: A Sketch," *Cultural Anthropology* 27, no. 3 (2012): 429.

42 George Marcus, "The Ambitions of Theory Work in the Production of Contemporary Anthropological Research," in *Theory Can Be More Than It Used To Be*, eds. Dominic Boyer, James D. Faubion and George E. Marcus (Ithaca, NY: Cornell University Press, 2015): 52, 55.

43 George Marcus, "Art (and anthropology) at the World Trade Organization: Chronicle of an Intervention," *Ethnos* 82, no. 5 (2017): 922, 923.

44 Dominic Boyer and Cymene Howe, "Portable Analytics and Lateral Theory," in *Theory Can Be More Than It Used To Be*, eds. Dominic Boyer, James D. Faubion, and George E. Marcus (Ithaca, NY: Cornell University Press, 2015), 17.

45 Edward Said, *Orientalism* (New York: Pantheon Books, 1978); Gayatri Chakrovorty Spivak, "Can the Subaltern Speak?" in *Marxism and the Interpretation of Culture*, eds. Cary Nelson and Lawrence Grossberg (London: Macmillian, 1988), 271–313.
46 Dipesh Chakarbarty, *Provincializing Europe: Postcolonial Thought and Historical Difference* (Princeton: Princeton University Press, 2000), 16.
47 Dominic Boyer and Cymene Howe, "Portable Analytics and Lateral Theory," in *Theory Can Be More Than It Used To Be*, eds. Dominic Boyer, James D. Faubion and George E. Marcus (Ithaca, NY: Cornell University Press, 2015), 18, 26–27.
48 Gillian Hart, "Relational Comparison Revisited: Marxist Postcolonial Geographies in Practice," *Progress in Human Geography* 42, no. 3 (2018): 371–394; see also Jennifer Robinson, "Comparative Urbanism: New Geographies and Cultures of Theorizing the Urban," *International Journal of Urban and Regional Research* 40, no. 1 (2016): 187–199; Ananya Roy, "What is Urban about Critical Urban Theory?" *Urban Geography* 37, no. 6 (2015): 810–823.
49 Shu-mei Shih, "World Studies and Relational Comparison," *PMLA* 130, no. 2 (2015): 436, original emphasis.
50 Shu-mei Shih, "Comparison as Relation," in *Comparison: Theories, Approaches, Uses*, eds. Rita Felski and Susan S. Friedman (Baltimore: The Johns Hopkins University Press, 2013), 79.
51 George Marcus, "The Ambitions of Theory Work in the Production of Contemporary Anthropological Research," in *Theory Can Be More Than It Used To Be*, eds. Dominic Boyer, James D. Faubion and George E. Marcus (Ithaca, NY: Cornell University Press, 2015): 48–64.
52 Michael Callon and Vololona Rabeharisoa, "Research 'in the Wild' and the Shaping of New Social Identities," *Technology and Society* 25, no. 2 (2003): 193–204.
53 Matthew Dillon, *Pilgrims and Pilgrimage in Ancient Greece* (New York: Routledge, 1997).
54 Sara Ahmed, "A Phenomenology of Whiteness," *Feminist Theory* 8, no. 2 (2007): 153, 156, 162.
55 Diana Taylor, *The Archive and the Repertoire: Performing Cultural Memory in the Americas* (Durham and London: Duke University Press, 2003).
56 See Alex Lazaridis Ferguson, "Improvising the Document," *Canadian Theatre Review* 143 (2010): 35–41.
57 Diana Taylor, "Save As," *On the Subject of Archives* 9, no. 1 & 2 (2012): n.p.; Diana Taylor, *The Archive and the Repertoire: Performing Cultural Memory in the Americas* (Durham and London: Duke University Press, 2003), 36, 21.
58 Shannon Jackson, *Social Works: Performing Art, Supporting Publics* (New York and London: Routledge, 2011), 42.
59 For example Linda Alcoff, "The Problem of Speaking for Others," *Cultural Critique* 20 (1992): 5–32; Teresa de Lauretis, *Technologies of Gender: Essays on Theory, Film, and Fiction* (Bloomington: Indiana University Press, 1987); Aiwha Ong, "Women Out of China: Traveling Tales and Traveling Theories in Post-Colonial Feminism," in *Women Writing Culture*, eds. Ruth Behar and Deborah Gordon (Berkeley: University of California Press, 1995), 350–372; Gayatri Chakrovorty Spivak, "Can the Subaltern Speak?" in *Marxism and the Interpretation of Culture*, eds. Cary Nelson and Lawrence Grossberg (London: Macmillan, 1988), 271–313.
60 Richa Nagar, "Hungry Translations: The World through Radical Vulnerability," *Antipode* early view (2018): 2.
61 Lauren Berlant, *Cruel Optimism* (Durham: Duke University Press, 2011).

62 Saidiya Hartman, *Scenes of Subjection: Terror, Slavery, and Self-Making in Nineteenth Century America* (Oxford: Oxford University Press, 1997); Saidiya Hartman, *Lose Your Mother: A Journey Along the Atlantic Slave Route* (New York: Farrar, Straus and Giroux, 2007); bell hooks, *Yearning: Race, Gender, and Cultural Politics* (Boston: South End Press, 1990); Chandra T. Mohanty, "Under Western Eyes: Feminist Scholarship and Colonial Discourses," *Boundary 2* 12, no. 3 (1986): 333–358; see also Carolyn Pedwell, *Affective Relations: The Transnational Politics of Empathy* (Basingstoke: Palgrave Macmillan, 2014).

63 Though we mention the Berlin production briefly in Chapter 1, we do not give this production much attention in this book, for the simple reason that we were not yet alert to the possibilities of recording the play's travels until it had established itself as a more seasoned and inventive traveller.

64 When she engages with theory outside her community, Nishnaabeg scholar Leanne Simpson asks herself a series of questions:

"Where does this theory come from? What is the context? How was it generated? Who generated it? What was their relationship to community and dominant power structures? What is my relationship to the theorist or their community or the context the theory was generated within? How is it useful within the context of my own people? Do we have a similar concept or theory? Can I use it in an ethical and appropriate way (my ethics and theirs) given the colonial context within which scholarship and publishing take place? What are the implications of citation, and do I have consent to take this intellectual thought and labor from a community I am not part of? Does this engagement replicate anti-Blackness? Colonialism? Heteropatriarchy? Transphobia?"

This line of questioning is, she suspects, one that Indigenous scholars do naturally. White scholars at home in universalism generally do not. But they should. (Leanne B. Simpson, *As We Have Always Done: Indigenous Freedom through Radical Resistance* (Minneapolis: University of Minnesota Press, 2017), 63).

65 Carol McGrahahan, "Theorizing Refusal: An Introduction," *Cultural Anthropology* 31, no. 3 (2016): 319; Carol McGrahahan, "Refusal and the Gift of Citizenship," *Cultural Anthropology* 31, no. 3 (2016): 335.

66 Leti Volpp, "The Indigenous as Alien," *UC Irvine Law Review* 5 (2015): 325.

1 Performing a research archive

Nanay is a site-responsive play created to tell the stories of Filipino domestic workers and their families in Canada and of Canadian families in need of care. It brings into performance a big national story about indentured servitude, the long-term marginalisation of a racialised community and inadequate state support of childcare and eldercare through a series of intimate monologues and other scenes. The process of creating the play began by identifying research transcripts that seemed rich enough to sustain a monologue: because the interview was detailed, a person emerged off the page, or their experiences brought an important perspective to the issue. In February 2007, we worked with Alex Ferguson for a week with three professional actors at Vancouver's Playwrights Theatre Centre editing the most promising interview material into monologues, experimenting with staging and the possibility of creating dialogues or scenes out of a composite of interviews. Very little of this material was used in the end, but documentation of the four or five scenes worked up during this workshop convinced producers at Vancouver's PuSh International Performing Arts Festival to include our play in their 2009 programme.

Our work began again in earnest in the spring of 2008 when we and a member of the PWC of BC (now designated as writers) began to work with a dramaturge, Martin Kinch. Alex Ferguson (now the director) took us (along with five professional actors, a stage manager, a scenographer, an additional set designer, lighting and costume designers, and three Filipino youth apprentices) through a two-week development workshop in July of that year, from which emerged *Nanay: A Testimonial Play*.

The play was developed and first performed in Vancouver's Downtown Eastside at Chapel Arts, located across from Oppenheimer Park, a place that reverberates with memories and historical incident. In the park, a community-carved totem pole commemorates the many Indigenous people who have lived and died in the neighbourhood,[1] and the documentary film, *The Battle of Oppenheimer Park*,[2] situates the present-day life in the park within a violent history of dispossession of the Musqueam First Nation from the unceded territory on which the park stands. An ill-maintained baseball diamond was – many years ago – home field for the famed Asahi Tigers, a

Japanese Canadian baseball team that was permanently disbanded during World War II, when members (and the entire Japanese Canadian community) were displaced from Vancouver and interned in camps.[3] The Kalayaan Centre, where the PWC of BC was located until 2010, lay on the other side of the park, a proximity that allowed for steady traffic between the theatre and the PWC of BC during the two-week development workshop in July 2008, and through periods of rehearsal and performance in January and February 2009.[4] Reopened in 2007 as a performance space, Chapel Arts was formerly a chapel and funeral home. We collectively created the nine scenes of the play in relation to the building's ambience and architectural form.[5]

A guided tour of *Nanay*: melancholic realism, disidentification and strangeness to oneself

The audience experienced the play in small groups, guided from room to room to hear Canadian employers and Filipino domestic workers speak about their lives; Canadian employers in the public areas upstairs where funeral receptions would have been held, domestic workers in the unheated, dank spaces below previously used for the delivery and care of corpses. At a PWC of BC community assessment after the event, one activist spoke of "*feel[ing]* the contrast – you know, the damp, the dark atmosphere downstairs, the cold and no [theatrical] lighting. So when you go up: the luxurious, you know, the well-appointed rooms. So it was really the best portrayal of the two solitudes: of the slavelike conditions, and the richness of the society that exploits these women."

Memories and loss run through the testimonial stories and performance spaces. Audiences saw domestic workers' monologues in one of the two sequences; one sequence began in the former embalming room, still fragrant with the scent of its previous function.[6] And while the Filipino Canadian actor Hazel Venzon told in her monologue – as she scrubbed and cleaned this place – of her great optimism, gratitude and joy leaving the Philippines to do domestic work, her departure is grounded in loss (Figure 1.1). She testifies to being abandoned by her husband when he left to do overseas contract work in the Middle East, which created the necessity of leaving her children, her mother, her father and her siblings in order to try her luck as a migrant worker, first in Hong Kong and then in Vancouver.[7] With a gruelling schedule of 16 performances on one of the performance days,[8] Hazel told us afterwards that she maintained her own stamina as an actor by silently dedicating each performance to a different female family member living in distant cities around the world: "this is for my mother, this is for my aunt..."

The audience was then guided into an adjacent garage-like room where the mechanism for raising coffins and corpses through the ceiling for public-viewing upstairs is still evident. The room was uninsulated and so cold that the actor was, of necessity, dressed in an overcoat. The actor, Lissa Neptuno, delivered the testimony of Joanne in a compressed space, wedged between

Figure 1.1 Ligaya speaks of leaving her children and "finding her luck" in Canada.

a small mounted cage and a calendar-grid drawn on the wall behind, upon which she charted her progress through the LCP (Figure 1.2). The director explains the rationale for the set: "the nanny had gone AWOL from an exploitative job situation that was no longer tolerable [...] so she was hiding out when she gave the original interview. I had hoped to imply vaguely that the scene was taking place in a safe house."[9] Joanne's angry monologue is about her treatment, as a professionally trained nurse, under the LCP, and it too is punctuated by loss, absence and defeat: of her husband and two children left in the Philippines; of being dismissed by her first employer; of the death of another elderly employer, which again left her temporarily unemployed; and of the revelation by the guide,[10] after her monologue is finished, that she failed, because of the succession of employers, to complete the programme within the required 36 months[11] and was obliged to return to the Philippines. She is absent. She is gone. We cannot alter her situation, even if we wish.

The audience was then moved to a narrow-darkened hallway to sit and listen to an eight-minute audio recording of mothers' and children's recollections of separation.[12] This is the only testimony delivered directly by

Figure 1.2 Joanne testifies.

domestic workers and their children (and not by professional actors). We hear only voices, symptomatic of their disembodied experiences of connection. The tearful voice of a domestic worker describing leaving her children is interrupted and overlain with the less impassioned voices of children recalling their memories of their mother coming and going, leaving and returning, and simply staying away for years at a time. At one moment, for instance, a mother speaks: "Four years and eight months. I never went home in-between these years. It's so I could have a better life in the future…." This is interrupted by a young woman's voice: "My mom would take care of me in ways that … she would send money and letters. But it's hard to connect when you're just sending letters, right? It takes a month before it gets to the Philippines." The mother's tearful voice resurfaces: "I did receive letters. Like if I sent them letters. I'd receive … up to two months, I'd receive an answer. And then I can answer that back after six months again. That's the only comfort I have there. You just cry every time you will miss your family

... It took me awhile, like six months, before I stopped crying when I get there. It was very hard." A young man, again in a more matter-of-fact tone, begins to speak over her voice: "She was writing letters, but we just get to see the letter. We don't really get to read it. We just see, 'Okay, she wrote letters.' But to us it's just like: 'Okay, yeah.' We were like ... we want to know how she is, but we didn't know how to show it kind of thing."[13]

The final scene in this sequence of domestic worker monologues was a space we called "the storied objects room." Domestic workers from the PWC of BC took charge creating the installation: it was a model bedroom assembled from memories of their rooms (Figure 1.3). An exact replica of a domestic worker's daily journal lay on the bedside table, inspirational and devotional religious passages interspersed with an unrelenting schedule of domestic duties. Christian iconography – images of Pope John Paul and a rosary draped on a small figurine of Christ – was carefully arranged in the small space. A small, ancient black-and-white television was turned on, positioned on a dresser at the end of her bed. (Oblivious of trespass, when a hockey game was on during one performance, some audience members perched on the end of the bed to watch it.) Around the edges of the room were framed, hand-written and drawn descriptions of different domestic workers' Canadian bedrooms (Figure 1.4), and on one wall, we installed a large collage of actual letters, cards and photographs sent between family members in the Philippines and Canada.[14]

As a tangible fully sensory container of memory, the bedroom was richly evocative for many who attended the play. Glecy, whose description of her Canadian bedroom as a domestic worker informed the piece, said that even though "I shared some of my experience there, I am still wounded [when I saw it] and I could still... It's a flashback in my memory when I hear [an audio recording of] the water splashing down [through pipes in the wall].

Figure 1.3 A model bedroom.

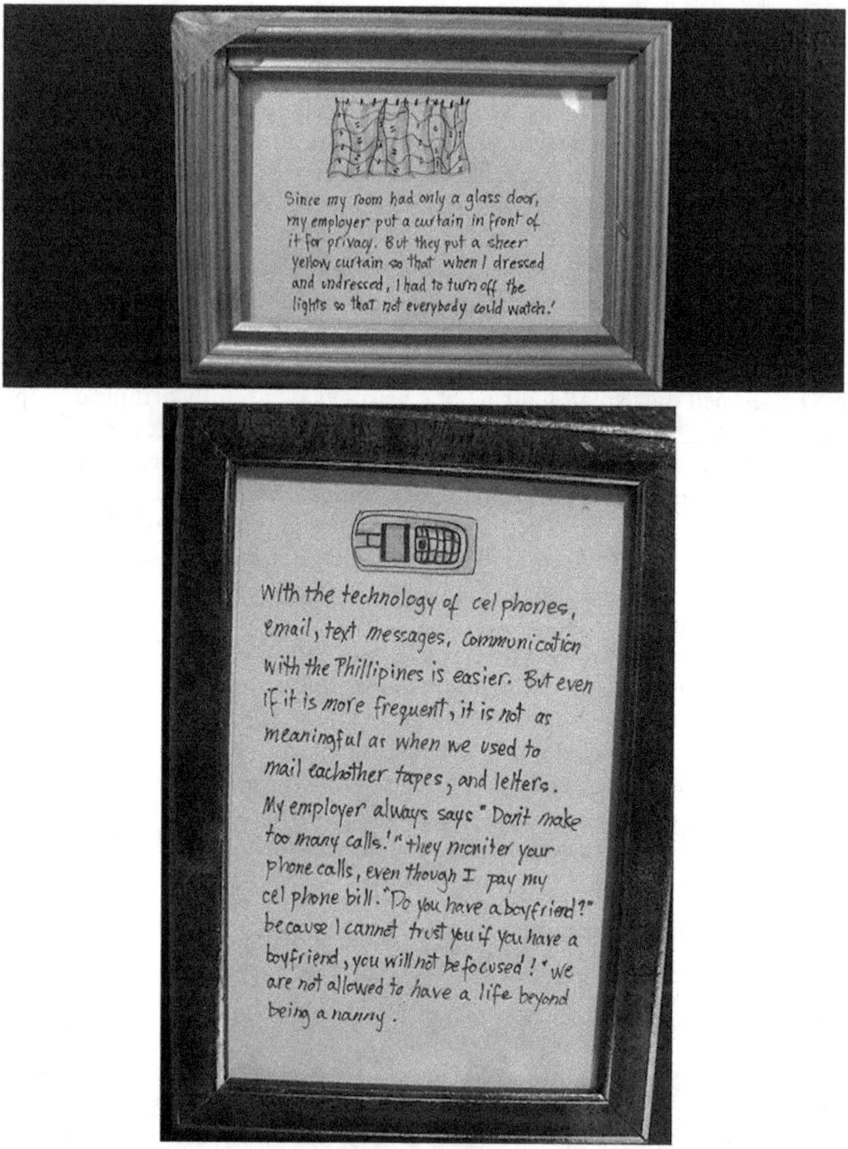

Figure 1.4 Framed memories.

Oh, it reminds me of how noisy my room was." From Coretta, who was not part of constructing the room, "like the bedroom ... I really can't control my tears. Because at the moment I can really feel the environment, the temperature, the smell of the room, as well as the materials around the room." Even those who had no direct memory of such a room felt moved by it.

Leah, a PWC of BC organiser whose son was born in Canada, spoke of its impact on her son: "But my son said that for him the one that impacted him the most was going into that bedroom because he felt it was so heavy. Like you couldn't breathe [because it felt so claustrophobic and the air was very stale], and it was quiet but not at the same time because of all the noise [of simulated water running through pipes in the wall] and the air. And the darkness. So it made a lot of the stories we hear tangible and sort of ... you are stepping into their space." Non-Filipino audience members wrote on an audience feedback survey of "a dreaded sense of hopelessness looking through the nanny's room" and that "the little bedroom with all the photos, letters and the like really affected me." "The replica bedroom was the most touching," wrote another; "the little room montage [of letters, pictures, and postcards] was so very sad." Before proceeding upstairs to hear testimony from Canadian employers, we pause to consider what kinds of ethical situations we sought to generate with these domestic worker monologues and take some measure of how they were received.

One reading is that these spaces of relics and embodied monologue set in motion what Ian Baucom has identified as a distinctive "knowledge grammar" for apprehending the world.[15] Baucom draws a distinction between what he calls speculative and melancholic realism, arguing that each operates in a distinctive epistemological register to create different kinds of observers. In what is undoubtedly a rough caricature of social science, Baucom argues that scientific facts are constructed in relation to the population or aggregate, and through a process of abstraction: by being emptied of local significance and networked into a system of thought. Melancholic realism operates differently, mimicking the process of melancholy. Within psychoanalytic theory, melancholy is understood as a process of refusing to mourn and relinquish what has been lost or to substitute it with other attachments. What has been lost is swallowed and preserved, and the subject constructs a "cryptic vault" of memory.[16] A melancholic insistence on the non-exchangable singularity of what has been lost has implications for how it can be expressed. In particular, because any representation is a form of substitution, representation itself is suspect, and melancholic realism thus operates through a "mode of reference that aims to pass itself off not as a representation of the lost thing but as that lost thing itself."[17] Consider also John Tagg's point that framing realism within melancholy "may help us think about the character of those practices of representation that will not give way to the demand for efficient communication but resist the arrival of meaning, while mourning a real that does not lend itself to representation."[18] That is, it might slow the witness down to prevent easy absorption.

The theatrical experience that we created is arguably working within the terms of melancholic realism, and the embodied, performative spaces of theatre are particularly powerful vehicles for it. The ghostly remains of Joanne are brought back to life: the organiser at the PWC of BC who did this interview spoke of the visceral impact of "seeing it alive again." In a community assessment organised by the PWC of BC after the Vancouver production, she spoke of the force of seeing this interview transformed into a theatrical performance:

> I've been involved [for many years] with doing community-based research, which involves documenting and interviewing women or nurses who come through the Centre to share their migration experience. So Sheila and I conducted one of the interviews [used to create Joanne's monologue]. And we didn't expect it to be part of the play. And I guess it was done four or five years ago. [Crying] And it was just seeing it alive again. You know, they were done and it's part of our daily organizing and documenting our history as a community. But seeing it presented live in the cultural way, it just struck in a completely different way that I didn't expect.

The darkened model bedroom embalms objects – an actual diary, the very letters and cards sent between mothers and children – in a tomb of memory. The domestic workers' and children's testimonies, each performed by a different Filipino actor in a distinctive setting, are constructed to incline the audience to the melancholic fact, to attach themselves and bear witness to the distinctiveness of migrant workers' experiences. Each domestic worker's look, tone, sensibility, manner of moving and speaking, mode of storytelling, and engaging the audience are distinct, not easily compressed into a representation of typicality. Encountering domestic workers' testimonies in a small group of 12, in intimate spaces where the domestic worker addresses you directly, for instance, asking you – that is you – to move in order that she can carry on her work as she tells you her story,[19] forces you – that is you – to engage her as a person, as a sensible and singular human being (Figure 1.5). To witness injury, Baucom argues, is precisely to attach to the singular, the historically particular, the unverifiable. It is, he argues, an entirely different

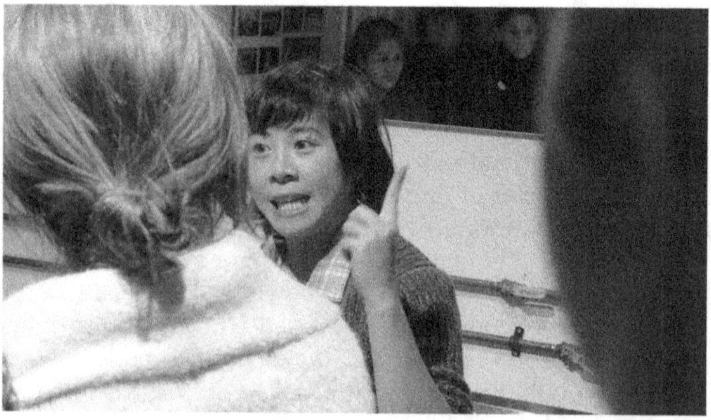

Figure 1.5 Ligaya tells her story close up.

mode of observation than is produced by speculative realism, which he takes to be the language of social science. The latter, he argues, encourages disinterested spectatorship of the universal, the exemplary or the theatrical fact. He argues that different politics take shape around these two observers: on the one hand, an interested melancholic politics of witnessing, which involves a tenacious, committed holding to an event; on the other, an impartial, disinterested, liberal universalising politics of abstract human rights. In Baucom's terms, the former is "cosmopolitan interestedness," and the latter is "liberal cosmopolitanism."[20]

We posed a choice between the two epistemological registers in the final scene of the play, for which all 50 members of the audience were reassembled. In the one constructed, non-verbatim monologue in the play, a representative of Citizenship and Immigration Canada (CIC) enumerated the costs and benefits of the LCP, harnessing a barrage of statistics to establish the balance in favour of the programme. He ended his monologue by introducing Michelle, a child of a mother who came through the LCP. Michelle (performed in Vancouver by Melissa Dionisio) was positioned at some distance from him, at the other end of a 20-foot table. Her testimony to family discord, depression and lowered educational and occupation aspirations in Canada relative to the Philippines[21] was often flat in emotion, and we are made to understand that this tone emerges from her migration experience: the loss of her friends and future in the Philippines and an absence of feeling for her mother in Vancouver. Giving the last word to Michelle, we as playwrights declared ourselves attached to her melancholic testimony of loss.

There are familiar risks that attend this attachment: risks of entombing Filipina women in their victimhood and of binding spectators too closely to Filipinas' injury and loss, thus instantiating the liberal subject of compassionate sentimentality. It replays what Martin Manalansan has termed a tragic linearity of victimhood attached to Filipinas.[22] Martin Kinch, our dramaturge, was dogged in his advice to counter these tendencies by introducing more positive testimony from Filipina domestic workers. Unfortunately, the scene of more positive testimony created in order to heed his advice was cut. We also worked with the monologue of Ligaya (a pseudonym chosen by the domestic worker on whose words the monologue was based, which translates as joy, happiness and pleasure) to emphasise her agency. We mined her interview transcripts for lines such as these:

> But I have a fighting spirit, something like that. I'm the eldest. I'm the rebellious one. And I think that is one of the qualities that gives me the spirit, the spirit to be courageous. Like look, I've been through a lot in life, and still I face it squarely. I'm not scared to be, you know, for being poor. And I went abroad. I left my children behind. I told myself I have to win all the battles. I have to fight them squarely and face every problem eye-to-eye. And yeah, I'm very thankful that my mom taught me this. With all these challenges, I'm a survivor.

Concerns about liberal sentimentality nonetheless remain. There are three features of *Nanay* that possibly disrupt the tendency to sentimentality, the *first* tied to the verbatim or documentary genre. The hybrid nature of documentary theatre encouraged audience members to wrestle actively and productively within the two epistemological registers that Baucom identifies. Documentary theatre, Ben-Zvi argues, is a "special type of double agent,"[23] deploying claims of truthfulness – in this case, actual social science truthfulness – while simultaneously using the aesthetic devices and embodied presence of theatrical performance. We can hear one audience member explicitly weighing melancholic and speculative facts in a talkback session:

> I found the stories of Ligaya [in the kitchen] and Joanne at the beginning [of the play] exceedingly powerful. It seemed to me that they dominated the whole play. My question is: how representative are they? It's hard for me to imagine how things could be much worse than they were for Joanne, up there in Whistler with three kids, working around the clock. But maybe it can get even worse than that. On the other hand, I would have thought it could get a good deal better, too: that there would be cases where families really do love their caregivers and where things are working out well. So these are two enormously powerful introductory stories. My question is about their representativeness.

This question – unsolicited – was repeated in almost every talkback and in a good proportion of the audience surveys.[24] But rather than dismissing this type of speculative realism, as Baucom does, as the mindset of the actuarial accountant, we might consider what each type of realism brings to the other. Arguably, one brings us close in order to experience the exposure and vulnerability of domestic workers; the other provides some distance necessary for political judgement.[25] We note, with some relief, that *Nanay* received awards for more than its emotionality as the "Saddest" play at the festival; the Children's Choice judges also considered it to be the "Most Realistic" and the "Most Interesting" play in the PuSh festival. And we take it as more than chance that the Most Interesting trophy dwarfed the other two that the play received; in its height and exorbitance of feathers, pompoms and beads (Figure 1.6). We hold these three awards in tension: saddest (affect), most realistic (realism) and most interesting (thought), to approach the politics and possibilities of *Nanay*.

A productive oscillation between interest and emotion, distance and proximity, judgement and its suspension, comes from a *second* source as well: the intercultural nature of *Nanay*, which forced non-Tagalog-speaking spectators in particular to recognise their inability to fully apprehend the play. We introduced the ambiguity of intercultural translation in the first instance through the naming of the play. A Tagalog word for mother, *nanay*, is almost intelligible to the English speaker as the word "nanny." It thus signals to English speakers that they both know and do not know the content

Figure 1.6 Children's Choice Awards.

of the play, or that they may not know what they think they know. It slightly shifts the ground of linguistic competence.[26] Beyond the name, the play was an intercultural *event*, and this fact fundamentally altered its meaning and reception. An audience survey taken at the Vancouver performances revealed the mixed nature of the crowd: 119 audience members identified themselves as domestic workers or family members, even more (184) had been employers or knew employers of domestic workers, and 255 had no relationship with anyone who had direct experience with the LCP (they were there for the theatrical experience). The PWC of BC ensured that at least five domestic workers or family members attended each performance, and the presence of so many Filipino domestic workers and family members as fellow travellers both drew Canadian audience members close to domestic workers' stories and kept them at a productive distance. One audience member commented, "I found it really moving to see the show, as someone without direct experience of the program, being able to see it with people who had had direct experience with the program. I found it extremely moving."[27] But equally, another wrote in the audience survey about the significance of gaps in cultural understanding: "[What resonated?] Seeing women in the audience crying at fragments of conversation in the [sound room of audio recordings of mother and children describing their periods of separation] – that I couldn't entirely follow/get. Seeing phili audience nod during Canadian gov't testimony."

There is a peculiar scopic economy in the performing arts as compared to a written archive or other visual arts that heightens this attentiveness to intercultural reception. This is especially so in a performance such as *Nanay*, with the absence of theatrical lighting in the domestic workers' scenes, and an intimate proximity between actors and audience, and among audience members themselves. Filipino domestic workers read Canadian audience members for their reactions. Ate Letty recounted, "I'm pretty sure some Canadians really don't know what is happening. Really don't know the reality of the live-in caregivers. I know that some of them reacted at the play because I was there, too." From Glecy,

> I have these feelings [at the play]: maybe they [non-Filipino Canadians] will get upset or how can they accept the message or the contents of the show? But my observation of other audience [members] [...] when we go first into the sound room I could feel that the majority of my group is really amazed. They were saying: how could they get the person to share their story? How do you ask individuals to speak out? I could hear their whisperings, like "Oh, this is not an easy job." For me, I can say this is not an easy job: it's hard to overcome the deskilling or your self-esteem being lost.

A *third* break from liberal sentimentality potentially comes not from the staging of the domestic worker scenes but through the employer scenes upstairs. And so let us climb the elegant stairway in Chapel Arts to encounter these scenes. For these parts of the play, a larger audience of 26 was reassembled and placed in more conventional theatrical bleacher seating, to view from a stable and static vantage point three-employer scenes.

Staging the employer scenes was the source of the greatest conflict among those who created the play.[28] The director, Alex Ferguson, was challenged to find the complexity of many employers' characters or to empathise with their circumstances. A short excerpt from an employer scene may convey a sense of the challenge:

RICHARD: When Stephen was six months old, we chose a Filipino nanny because we heard that they were very caring for the very young ones. So we basically only interviewed Filipino nannies.

STEPHANIE: We found out about Marlena from a friend of ours. How we worked it out was like this: we had two bedrooms upstairs and one room that we used as an office. So we sacrificed that. In that information booklet, it told what a live-in caregiver is entitled to have. And it was a room with sleeping arrangements, and a lock on the door. Although no one's ever locked the door.

RICHARD: And then we also gave her separate bathroom facilities. And she didn't need a separate phone, but we gave her one. We gave her a TV, a desk, an answering machine. It's different than working in Singapore or Hong Kong. Marlena told us stories of where the nannies were sleeping. It wasn't a pretty scene.

STEPHANIE: They're treated like second-class citizens in other countries!

RICHARD: At first she wanted to call us "Madam" and "Sir"! But we said, "Whooah, wait a minute." I think she was kind of taken aback by that. And we said to her, "That's not the Canadian way."

STEPHANIE: More than anything, we've become friends.
RICHARD: Yeah, we wanted to break the ice.

This exchange accommodates a comedic parody, an opportunity that the director took up through his choice of set (a bedroom overlooking a picturesque scene of gentrified downtown Vancouver), selection of costumes (plush bathrobes followed by a quick change into formal evening wear), and emphasis of particular words, tone, gestures and gloating glances between husband and wife (Figure 1.7). And yet, as Judith Butler observes, the risk in parodying employers is that "condemnation, denunciation, excoriation works as quick ways to posit an ontological difference between judge and judged" and can have the effect of foreclosing a fuller knowledge of oneself.[29]

For the other two scenes, we sought the most sympathetic employer stories that we could find in the existing archive of research material, in order to force the audience into a nuanced relationship with Canadian employers. Against the advice of the dramaturge, who felt the risks to friendship were too high – a concern that raises other ethical concerns – we conducted more interviews with friends and acquaintances in our search for less easily stereotyped employers.[30] In one of the two scenes we eventually included, two white middle-class women sit across from each other in a kitchen, elaborating their futile efforts to find childcare in Canada (Figure 1.8). Taken from two separate research transcripts, it is staged as two concurrent monologues that thread into each other, with the women repeating, repeating, repeating the same experiences of Canadian caregivers coming and going, being hired and fired, or just failing to show up. Each extremely stressed by the unreliability of Canadian home-based caregivers, they come to different ways of alleviating their stress: one hires a Filipino nanny through the LCP, the other quits her job.

Figure 1.7 A Canadian couple speaking about their nanny.

Figure 1.8 Two Canadian women at wits' end.

It is the third employer scene that comes closest to the testimonial genre of the domestic worker monologues and encourages the audience to engage with the complexity of her situation. In this scene, a middle-class academic tells of the pain and difficulty of arranging 24-hour care for her mother, who has had a stroke and is coping with Parkinson's disease and wants to stay at her home: "When she came home from three months in the rehab hospital [after the stroke], I woke up one morning and she was on the floor and she couldn't get up. She said she'd been there for five hours and she's freezing cold because the air conditioning was on and she didn't have a blanket on her. I just started to cry. It was awful. And after that happened she finally resigned herself to the fact that she did need someone all of the time."

The woman who testifies to her experience articulates a critical perspective on the work of social reproduction: "My mother doesn't understand why … she doesn't really see it as a job, taking her to the bathroom. She just thinks of it as like something you would do for anybody. If you just happened to live there, you would do it. She doesn't see it as wage-gaining work. It's like it doesn't compute to her that people have lives and they have other things to do. That this is actually labour." Anticipating the criticism that the relationship is exploitative, and recognising that she pays the Filipino domestic worker who has come to Canada under the LCP less than she pays the Portuguese Canadian housekeeper who has cared for her mother during the day for many years, she wrestles with the ethics of the situation and her economic constraints, worrying and rationalising, weighing and balancing her political commitments against her own economic resources, and her love of and uncompromised allegiance to her mother. The testimony anticipates her loss, her mother's death, and documents her efforts to make her mother as comfortable as possible as her body deteriorates towards this end. It is difficult not to sympathise with this loss and the trade-offs that she makes.[31] To witness – truly witness – her testimony is to enter into a complicit relationship with the LCP and to entertain the possibility that you too could become a domestic worker's employer.

These employer scenes proliferated attachments to people positioned differently to the issue. In a community forum after the Vancouver performances, a Filipino-Canadian activist, father of two young children, spoke of being "surprise[d] to like parts that I didn't think I would like. In particular, the two women in the kitchen. I don't know why it resonated so much, but I think it was really seeing the Canadian issues collide with the [Filipino] community's issues as two mothers, I guess they are single moms, I'm not sure, but just talking about their difficulty finding child care as a family. [My reaction] was really surprising to me." Attaching to the stories of *both* domestic workers and Canadian employers takes the play beyond the sentimental: absorbing multiple and divergent stories of loss forces critical thought and some hard political work. As one audience member put it, "They all sound real AND they also seem to contradict each others' stories. I felt compassion for these overstressed women/employers AND for the equally exploited overstressed workers."

Knowing oneself as a potential employer is possibly to know something new and unwanted about oneself. As Butler has argued, it is relationality, and not the (autonomous) moral subject, that grounds ethics.[32] Foreignness to one's self – a rupture in self-mastery and a sense of being acted upon by relations that one has not chosen – is key, in her assessment, to this ethical connection with another. By this schema, it is not one's moral superiority to Canadian employers who hire domestic workers that grounds ethics. It is being acted upon by relations that make the LCP a viable option for oneself – perhaps one of the only viable options given current circumstances of childcare and eldercare in Canada – that establishes the grounds for ethical and political responsibility towards Filipina domestic workers *and* Canadians who need to care for their parents and children. Being caught between compassion for Canadian employers and Filipino domestic workers also refocuses attention, we hope, away from the issue of bad or immoral employers to a larger story of racialised global inequality, the scandal of undervalued care-work in Canada and the export of maternal labour from the Philippines. How all of this might shape up as politics is the question to which we turn by entering the last scene of the play.

A performative space of politics

At the end of one of the Vancouver performances of *Nanay*, a friend of Caleb's (a "theatre person") came up to him in a state of bemusement. He said something like this: "I don't know what the fuck just happened in there. First this one woman starts asking the actor questions from the audience. And then a second asks her a question...." He was referring to the final scene of the play, when the child of the LCP speaks her monologue. In an effort to make transparent the constructed nature of the monologues, the director staged this particular scene closer to its original form as an interview. Seated among the audience, in role as researcher, Gerry prompted the young woman's monologue through a series of short questions. This evidently destabilised the boundary between performers and audience quite effectively because on this occasion an audience member filled a brief pause in the actor's monologue with the sincere question: "How do you get along with your mother now?" In an instant of astonishing good fortune, the next line answered the question.

The current attraction of documentary or verbatim theatre, some argue, is that it fills an important gap in the political landscape by feeding a public appetite for a more diverse set of stories and opinions than are delivered by conventional media, so as to stimulate more complex analysis and debate.[33] In short, it is an antidote to infotainment. While this may be, it is more than the content or information conveyed that is significant. Its potential lies, as well, in being a performative rather than didactic space. In this we follow Jacques Rancière's argument that the emancipatory potential of theatre rests not so much in the genre or subject matter of the work presented, but in the relationship of the audience to the material performed, the actors, the space and to one another.[34] Theatre's political potential rests in the opportunity it provides to blur the opposition between those who look and those who act (between passivity and the capacity to take action), and between those who are locked within their functions, roles and social identities and those who exist beyond them. The potential lies in being and not just professing an egalitarian space.[35] Theatre can be a distinctive public space where it is possible to *perform* (and not simply preach) equality.

Nanay can be interpreted as an egalitarian space in the first instance because it offered migrant domestic workers, in their precarious and slippery status as non-citizen residents, an opportunity to command the attention of a mainstream audience on an equal footing with Canadian employers and a representative of the Canadian government. But opportunities for democratic participation go far beyond this.

The equalising potential began with the location of the play, where many in the audience felt somewhat out of place. When we first selected the location of Chapel Arts, some of us had concerns that its marginal and highly stigmatised location in Vancouver's Downtown Eastside would discourage a middle-class audience from attending.[36] And although the venue was so close to the Kalayaan Centre, comments made by Filipino activist collaborators suggest that they too saw it as a foreign place: "Seeing the play in a completely mainstream setting – it gave us an idea of how much we have to do to continue the debate in the public arena"; "you attract a different … not too many people can go to elite venues. So bringing *Nanay* into [Chapel Arts], we are also reaching out to people who would not otherwise go to the Kalayaan Centre."

The structure of the play and the audience's mobility through it also disrupted a sense of authoritative control over interpretation and participation.

Filipinos and non-Filipinos, those with and without connections to the LCP, employers and domestic workers, theatre patrons and activists – these audience members were arbitrarily assigned at the beginning of the play to one of three groups that were informally guided through the monologue/scenes in different orders. Their mobility and informal sociability contrast to what Rancière characterises as the policing of audiences through a system of reserve seating, the audience positioned "like so many temporary owners of property."[37] The structure of the play is itself episodic and fragmented, with a multiplicity of entry points. Some encountered the employers before the domestic workers; others saw the show in reverse order. Domestic worker scenes were experienced in different sequences. There is no single narrative arc, suggesting, we hope, an open encounter and interpretation. At least some audience members noticed, "I enjoyed how piecemeal it was – to step in/out of different stories/perspectives added physical/material complexity to an already complex issue."

And although we used professional actors, neither actors nor audience members were locked into their roles. There was much speculation among audience members about the blurring of the boundaries, as well as active transgression of them. We collectively chose to use professional actors because our community collaborators were interested in developing skills and networks made possible by working with theatre professionals, and because those whose testimony was used could not appear in the play. The domestic workers did not have the luxury of time or money to leave children and employment for rehearsal and performance, and issues of privacy and consent were complicated in the case of the child of the domestic worker who revealed details about her father's girlfriends. In any case, the politics of "real" people performing their authenticity are as fraught as the danger of appropriation; in neither situation do power relations go away.[38] Although we appreciate the complexity of and are undecided about this argument, actors also open an interesting possibility for democratic politics precisely because they have no sanctioned authority to speak the words they speak, or at least none grounded in an identity, mirroring (or literally performing) the contingency of the democratic authority to speak.[39]

Audience members actively searched across the blurred line of acting and identity, fiction and fact, genuinely troubled by the indistinction. In audience surveys, they made comments such as "What is real? What isn't?"; "Are actors people who went thru LCP?/hired nannies?"; "The way they portray their roles were really really real"; "Do the actors or any of them have any reaction about the true stories of the women? How would they feel if they were in the story?"; and "When looking 'the' nanny's diary in the bedroom, I felt uncomfortable not knowing if it was a fictionalised diary or a real one." Especially for the Filipino actors, audience members tried to fix the actor in their character: "The philipina actresses were extremely powerful, genuine, and convincing – I believe the stories resonated more with them (family backgrounds) than it did for the Caucasian actors, who probably had no connection to this issue." And though we might (and do) read this as a tendency to naturalise migrant care-work in the body of the Filipina, the effect may also have led in the opposite direction. As on audience member put it, "The 'Philipina' part was overall more effective in troubling issues of representation – it wasn't clear if these women were actors or testifying at times (which catches the viewer in a lot of assumptions)."

Audience members themselves were destabilised in their role,[40] to the point that one young woman at one performance asked for clarification of the norms as they had developed within this particular production: "I have a question to ask everyone who watched the show. I didn't clap at all during any of the show. I wanted to clap. But – I don't know if it was the tone or mood – I didn't know how to express how grateful I was for them doing the show. I was wondering ... did anyone clap in any of the sessions? ... Is it okay if we clap?"

Because of the ambiguity of the distance between reality and fiction and the intimacy of actors and spectators, audience members periodically crossed the line from spectator to actor. A young audience member became part of the action, for instance, when, pressed against the wall, he inadvertently turned off all the lights in the cramped kitchen scene. The actor simply moved him, turned on the lights and scolded him (as a nanny might), telling him she preferred to work with the lights on.

But more than these periodic improvisational eruptions within the performance, the expectation of active participation was demanded by the fact that the talkback session flowed directly out of the performance without break. For all performances, most of the audience stayed for the talkback, about 50 people arranged to face one another in conversation (Figure 1.9). The actors, producer and director often slipped in without comment among the audience as equals (rather than experts) in a conversation that typically focused not so much on the play but on the economics, politics and (in)justice

Figure 1.9 Public forums.

of the LCP.[41] The talkbacks created the time and space for extraordinary public conversations between those who most likely would not otherwise speak or listen to one another: domestic workers spoke directly to employers, activists to government officials, activists to other activists aligned around different issues. In audience surveys, the talkback was described as "a very interesting space." "I liked the talkback and the permission it gives to the audience to ask questions and to hear more directly from the community." "[What resonated?] The talkback with the real women speaking"; "Appreciated the personal accounts in the talkback session especially"; "I loved the talkback between the employers and caregivers and their families. Destabilised the notion of audience. Moving, compelling."

Performance theorist Jill Dolan writes about the potential of performance to "inspire moments in which audience members feel themselves allied with each other, and with a broader, more capacious sense of the public, in which social discourse articulates the possible rather than insurmountable obstacles of human potential." She calls these moments of affective intensity among audience members "utopian performatives." They are moments of feeling or experience – not something received didactically as information – in which we get an inkling of "how powerful might be a world in which our commonalities would hail us over our differences." They "resurrect a belief or faith in the possibility of social change."[42] Such moments were palpable in some of the talkback sessions, and we seemed to witness personal transformation, right there, on the spot.

In one talkback, for instance, a white middle-class woman introduced herself as an employer: "We have a live-in nanny and she's wonderful." She recognised that her nanny was working beyond her contracted hours and posed the problem to the group: "My 7-year-old's job is to jump up from the dinner table and drag her from the kitchen… I wonder if there is some way, like some culturally appropriate way, that I can say that it's really okay to stop working at six. Because I want to be a good employer. Like the guilt associated with this show tonight: *oh my god* [laughing]." A member of the Filipino-Canadian Youth Alliance responded to this personal inquiry with a generalised argument about the deskilling of domestic workers, to which the woman answered acerbically: "It's very evident to me that she is skilled." Another audience member took a more congenial approach, first identifying herself as "half Filipino and half Canadian." My mom came here and of course I'm very Canadian. As my 16-year-old says, "The only Filipino thing about you is your mother." In her view, a "white empowerment model" was inappropriate: "I was raised by a Filipino mother to take pride in my work: whatever the job is, you keep a smile on your face and just motor on… What I mean is, help them to acculturate, help them to understand." An activist from the PWC of BC then shifted the discourse to one of rights and advocacy. This seemed to resonate with the woman who first posed her dilemma because she thanked the PWC of BC activist: "This is helpful. Like I guess I feel a little more comfortable because I asked her what her long-term goal was. I tried to sign her up for courses and pay for her time and that sort of thing."

This employer's concerns about being a good employer were forgotten when, with great emotion, a domestic worker spoke of her delayed understanding of Canada's family sponsorship regulations, which meant she could not sponsor her eldest son: "It was only when I finished the twenty-four months then I processed the papers, I read there: 'Oh my god, my eldest son cannot... He doesn't qualify.' We will be separated forever unless I go home [to the Philippines] and see him there. It's the only way." The Canadian employer once again spoke to the activists in the room: "Did someone say there are workshops in Tagalog because I would hate for this to happen to my nanny." From the domestic worker who could not sponsor her eldest son, "That's true." Another Filipina member added that her friend had suffered the same experience. The Canadian employer then erupted: "[When domestic workers] phone back home, do they say 'Canada's better' or do they say 'This really sucks' and tell everyone they know not to come to Canada because they've been lied to, or exploited, taken advantage of, the structure of thing sucks, 'I got tricked, my sick kid can't come when everyone else can.' So like, isn't there some kind of word of mouth or some kind of awareness? Journalistic coverage? Something?" Carlo Sayo, one of the talkback facilitators, turned the question to the domestic workers in the room: "Maybe that's something we can ask to some of the people who have been through the program. What do you tell your family back home?" In the space of 20 minutes, after hearing personal testimony from domestic workers themselves, the middle-class liberal employer moved from simply wanting to be a good employer to asserting that the "structure of the thing sucks."

Such moments of common feeling, unhinged from shared social identity, were deeply satisfying. Equally productive were moments of disagreement, when the possibility of translating across experiences was both refused and openly desired. During one performance, we, quietly chatting in the lobby, were surprised by a woman in tears rushing past us. She had left her group in the "objects room" and was headed to a bathroom for privacy. In the talkback, she identified herself as Colombian and explained what lay behind her reaction:

> I just wanted to know, do you ever have on this show actual Filipina workers? Because I felt, I felt really weird. I have been in the same kind of situation and I felt that looking at people [audience members] sitting on the bed [in the model bedroom], it was like an invasion of my privacy. And for me it was awful to see Canadians like, you know, for me it was a little bit offensive. During the whole play, I was wondering, like I felt like a coloured person watching the white people watching the play [...] Because when I have been in that situation and you're talking about me, that's really different. I felt that they were looking at my room. [...] I just felt that this play is for Canadians. It's not for Filipinos and it's not for immigrants.

Charlene Sayo, a Filipino-Canadian representative from the PWC of BC who had been involved in the play from its inception and was co-facilitating the talkback, countered,

> Yes, there were Filipino people that were involved, because the script was taken from interviews from Filipino domestic workers and there was consultation with members of the PWC of BC and SIK-LAB [another Filipino migrant organization]. In particular, in that room we did have domestic workers involved here, with their time and by donating their things and setting it up [...] In a very hands-on kind of way, they would say, 'Don't put that here. This has to go there. This has to go on the calendar [...]' For the women who participated, it was for them very liberating and empowering in the sense that people wanted to hear their stories and this was an opportunity to tell them.

Attempting to reengage the women's concerns, we added, "But absolutely people should go through that room in a very complicated way. I guess what you're saying is that you didn't witness this complicated relationship with the room."

It was almost 15 minutes, after the Colombian woman had slipped out of the room, that another audience member, this time white, male, in his early thirties, returned to her comments. He began by introducing himself: "I'm [first name]. I'm a theatergoer, theatre artist, and landscaper and stuff." He proceeded to analyse the bedroom from this perspective:

> I thought it was a very interesting choice to allow us, to invite us openly, to encourage us to go into that bedroom and to look at that bed and to sit on that bed. And how you carpeted the platform [on which the bed is placed]. Very interesting choice and definitely a very complicated interaction. As an audience member and as a theatregoer I am trained – and I think that is an appropriate term – to engage with a piece. And obviously I engage with it from my own background. And I think it's fascinating and I think it's a great choice and I'm really interested in dialoguing with that.

He then returned to the Colombian women's comments:

> Because personally, the reaction of the women who's left now – and unfortunately I didn't have the courage to speak up earlier but I wish she was still here – is a really interesting perspective for me because I had the reaction, 'That's wrong [to enter this space] but this is a theatre piece and therefore I can do it.' What I'm interested in now is that it evoked such a strong reaction. Because she could be speaking to me directly. I'm interested in opening a dialogue about that because, how do you [know]? I mean I don't know. I mean I'm only me as a person

> from Canada. And how do you? So I just want to open a dialogue because I don't know if it was a piece for white people. It feels to me like it was piece… I don't know. So that's a question for you guys. [...] I would like to engage with that anger more. Because it's quite rare. And I don't mean from a masochistic point of view. Because I think here in Canada, certainly here in Vancouver, we exist within a really politically correct environment, which is for a very good reason but it is also an environment that stifles conversation and dialogue and the ability for people to engage in that way. I don't know if what I'm saying is massively offending people. But I'd be really interested to hear more of that angle and more of that perspective.

"She could be speaking to me directly." The talkback evidently created a time and space to take the risk of sounding angry, offensive or foolish, that is, to take the risk of encounter beyond or at least across established identifications. It created a time and space for an immigrant woman to level the accusation that the play is offensive and exclusionary, and "a theatregoer, theatre artist, and landscaper and stuff" to invite a conversation that would allow him to see the work differently, to disturb his "trained" eye, to better understand whether and how his world is constructed "for white people." In short, in Rancière's phrasing, this talkback created a time and space to redistribute the sensible: reconfigure what can be seen, heard, felt and thought.

One of the most interesting talkbacks occurred when three representatives of the Philippine embassy attended the play. This talkback was itself a performance – an actual demonstration – of the discursive working of state power. As a liminal space, the talkback brought into conversation people who would not usually have or take the opportunity to speak to one another. Embassy representatives dealt with this situation using very effectively two counter political strategies: privatising and minimising the claims to injury made by migrant domestic workers and, by extension, the play. Rancière claims of state authority that "rather than solicit a submissive subjective recognition or response, [those in power] dismantle political stages by telling would-be spectators that there is nothing to watch. They point out 'the obviousness of what there is, or rather, of what there isn't: Move along! There is nothing to see here!'"[43] The exchanges between one of the facilitators (the representative of the PWC of BC) and the embassy representatives took this form. We restage one here:

PWC REP: In the Philippines we are forced to leave our country because we don't have a future. There is no job in the Philippines. [...] the government is targeting 1 million workers going abroad [for the remittances].[...]

EMBASSY PERSON 1: I am [...] from the Philippine embassy. You could also confirm this by searching for the information online; it has never been the policy of the Philippines government to send domestic workers abroad. In fact it is perhaps a consequence of the fact that they would like to look for the so-called greener pastures. But what we have always been trying to do is to keep people home where they belong.[...] So it's not the policy. I just wanted to clarify that.

PWC REP: I know it's not the policy, but before we export fruits, now we export humans. It's not a matter of choice. [...] I didn't want to leave the Philippines [...] but I don't have a choice. I need to support my family in the Philippines. [...] So it is not a matter of choice. We are forced to leave the country.

EMBASSY PERSON 2: If I may add, it just shows that people are free to leave if they want to leave. The official migration policy started roughly in the 70s. [...] It was just a temporary measure at the time when there was an oil crisis but somehow it seemed to gain some roots. But officially now, for the past five years, it has not been the official policy of the Philippines to send labor abroad because the social costs far outweigh the remittances if we add them up.

ALEX FERGUSON: Did you say for the last five years it hasn't been a policy? Because I was under the impression that there used to be a policy.

EMBASSY PERSON 2: That was the brain-child of the late foreign affairs secretary, Blas Ople. Somehow in the Labor Department it became roughly there. But ever since five years ago or so it became very clear in the Department of Foreign Affairs, which has been sending reports to the Home Office to stop this employment abroad, especially in the Middle East [...]. I mean Canada appears to be good. But the official stance from the Labor Department is to continue sending. So somehow there is a dichotomy [...]. But officially the Philippines does not promote labour as its export. Never.

In this short exchange, embassy officials use a discourse of individual choice to privatise and individualise the situation. The statement made by the representative of the PWC of BC about the labour export policy is refuted ("it has never been the policy"), but with a question from the director, Alex Ferguson, it is localised in space and time: in the seventies and now in the Department of Labor. The situation in Canada appears to be good, suggesting that there is not much to look at here. The PWC of BC representative's expression of her experience of forced migration is ignored: "If I made add, it just shows that people are free to leave." She was later further disqualified when the second embassy person answered a question from the audience about the length of separation of mothers from their children: "If you take a country like Saudi Arabia... An accountant can go home three, four times a year. An ordinary nanny could go home after two years, or three... If you belong to a lower class, you suffer more." This social order, including a hierarchy of rights to family life, is taken as a given: it is unremarkable.

This open public forum nonetheless created the opportunity for direct commentary and confrontation. For instance, activists from a Europe-wide migrant domestic worker advocacy organisation, RESPECT, countered (one member translating for another from Spanish to English): "For her [her fellow activist for whom she was translating], she feels that your discourse as a representative of the state is kind of messed up. All the things that you are doing for them. All the things that you receive from these workers. We're not sure what help is coming from the state. There's a mixture of discourse." A good number of audience members, beyond those representing migrant domestic workers, challenged the embassy representatives in a heated talkback conversation that went on for well over an hour.[44] But while the conversation was exhilarating, it also allowed the embassy representatives the opportunity to explicate more fully the official discourse. One of the embassy representatives ended the talkback with a long impassioned speech, delivered first in Spanish and then in English.

EMBASSY PERSON: And to the gentleman who has been asking that you must see that the social cost is so great. Of course we do. Canada did not need to make this study or this play for us to know the problem that we have. It is everywhere. [*Move along*] It is something that is even in the papers and in countless studies before we came today, that the problems and the social costs of migration are stupendous. We end up educating people for a career that they can never practice perhaps because there is no job waiting for them. But these are things that we are already working towards solving. [*Move along*] [...] And the lady [PWC representative] said that so many Filipinas approach the Philippine Women's desk in Canada. In fact I'm sure the consulate and embassy in Canada are using you as a partner in reaching out to the Filipinos. [*Move along*] There may be cases [...] of some abuse. [...] That is the job of the [Philippine] labor attaché. And if he is not doing his job well, it is also your job to take him to task to answer for that. [*It is your responsibility.*] [...] I can also share with you the experience of a young man: I think he topped either the law or medical board exams back in the Philippines and he was very proud to say that his mother is a domestic worker. There are many stories out there about children who know exactly what their parents give up to give them the kind of future they are enjoying now. There are. It is not always that you have an ungrateful child who has ended up having wayward ways. You also have some happy endings. *Your testimonials are selective.*]

PRATT: Except in that case it came at the cost of family separation.

EMBASSY PERSON: Oh yes, of course.

Oh, yes, of course: hardly a decisive victory for either *Nanay* or critics of the LCP. But what we can claim for *Nanay* is that it created a public space in which it is taken as foundational that all participants have an equal capacity to speak and an equal right to be heard. This is, as Rancière notes, "a reasonable-unreasonable idea." It is a reasonable idea that is unreasonable if one considers the way that societies are actually structured. *Nanay* creates public spaces that not only authorise but perform the rights to and capacity for speech of those with uncertain or low status. Such a performance, Rancière argues, is "always both argument and opening a world where [such] argument can be received and have impact."[45] It is, in other words, a performative argument, one that posits the possibility of a common world of argumentation.

Outside the theatre

Criticisms of Rancière's theory of politics nonetheless trouble an easy celebration of the political possibilities of *Nanay*. Jodie Dean argues that moments of staging equality have little effect beyond producing a sense of pleasure and satisfaction among those who stage them. This is political visibility, in her view, without the risk of politics.[46] Peter Hallward delivers a similar criticism when he argues that Rancière encourages us to do little more than "play at" politics. "Once the stage is struck, little or nothing remains."[47] And while these are criticisms of Rancière's general theory of politics, they raise serious questions about what can be claimed for sporadic moments of egalitarian discussion, improvisation and contestation in the space of a theatre.

Performance theorist Jill Dolan has cautioned against measuring the success of theatrical performance by its impact on a world outside the theatre. "Let it live where it does its work best," she argues, "at the theater or in moments of consciously constructed performance." "The utopian performative, by its very nature, can't translate into a program for social action, because it is most effective as a *feeling*. … Perhaps such intensity of *feeling* is politics enough for utopian performatives." Such politics live in the desire "to feel the potential of elsewhere: the 'not yet' and 'not here.'"[48] As embodied, interactive events that take place in concrete spaces in real time, such performances are typically concerned, Tom Burvill argues, not so much "with ideological as with affective transformation in the audiences and constitut[e] a performance of ethics in themselves; they [are] forms of ethical practice."[49] Theatre can create the occasion to model a different, more vulnerable, egalitarian way of relating to others. Most of our lives, Lauren Berlant points out, occur in "modes of lower case drama, as we follow out pulsations of habituated patterning that makes possible getting through the day." Upper-case drama potentially creates a little world for a short period of time – a space-time – to "slow things down and to gather things up, to find things out and to wonder and ponder. 'What is going on.'" On occasion it might even constitute "the event of feeling historical in the present."[50]

We hold hopes, of course, that the space of the theatre and the event of feeling historical can bleed into and transform everyday life. Some audience members at *Nanay* did make productive connections between the play and their own lives outside the theatre. They created, to use Rancière's phrasing, "their own poems with the poem that [was] performed in front of them."[51] We found these scribbled on audience surveys:

> I have known nannies through my work as a landscaper.
> We often communicate with the employer through them.
> They never let on for a second what is going on with them...
> Their situation.
>
> I really would love to have pre-service teachers come to watch this.
> I speak with teachers sometimes
> and occasionally there is this idea
> that parents who are more involved in school somehow care more.
>
> Thinking back
> I realize now how little I truly knew of our nannies' isolation
> and aloneness.
> Although we did our very best to include them in our family life
> they were so far away from home
> in such a different place
> I doubt they ever truly felt at home.[52]

It is difficult to trace where these feelings and these connections go once the show is over and the stage is struck. We do know that the fusion and confusion of art, testimonial, research, and activism proliferated audiences and multiplied opportunities for publicity. The play was reviewed in popular media as theatre, but more often it created opportunities to publicise and more widely disseminate our research; it spawned, for instance, full-length articles on the LCP in two widely distributed community newspapers. It offered opportunities for development and learning among our Filipino community collaborators: in the Vancouver production, Filipino activists were introduced to senior cultural funders, and there were three paid internships for members of the PWC of BC and the Filipino-Canadian Youth Alliance, as assistant stage manager and assistant directors. After seeing the play, a local government councillor posted a long review on his blog, noting that migrant domestic workers are "paying the price for our failure to create decent care for our children and seniors."[53] The director received the following email soon after the Vancouver performance: "I loved *Nanay* because it was completely not what I was expecting... I also had a great group with me, which certainly made the talkback more intense. Even afterwards, different advocacy group members were shaking hands and planning meetings with one another, and if that's not live theatre...."

Advocacy groups and audience members found connections over the need for national childcare and healthcare for seniors. One PWC of BC member spoke of the play as giving her "a space to try to understand how we can bridge those two groups [domestic workers and employers] while appreciating the distinction between the classes, between races, between genders."[54] A less inspiring but fascinating instance of chance connections occurred only days after the play closed in Vancouver, when the co-producers received an email from a nanny agent located across the country accusing us of slandering their business: this agency's website was one of many bookmarked on computers set up in the foyer for audience members to peruse while waiting for the performance to begin.

The HAU brought *Nanay* (and us) to Berlin in June 2009, where it was performed as part of the Your Nanny Hates You! festival on the family. The HAU is a much grander venue than Chapel Arts, and we staged the play to creep around the interstitial and back spaces of the theatre. The tone and meaning of the play were very different in this new context, in large measure because the voluble presence of Filipino domestic workers in the audience was lacking. But something else emerged in their place. From Vancouver, we had contacted RESPECT, a migrant domestic worker network that organises mostly Latino migrant domestic workers in Berlin. Not only was the representative of the PWC of BC, who travelled with the play to Berlin, able to connect with this network, but RESPECT arranged to come to a number our performances, which they utterly transformed by filling them with the insistent hum of the whispered translation of English into Spanish. In doing so, they put *Nanay* and our research more fully and differently into the world.

And yet

A number of reactions to *Nanay* nonetheless pressed at the limits of our research archive and the challenges of reaching towards representational strategies that do other than repeat the violences of the past and present.[55] Consider the following three sets of reactions.

One: in an effort to give the audience a short break from performed testimony and the monologue heavy play, the accomplished shadow puppeteer, Tamara Unroe, was commissioned by the director to create a shadow puppet scene. The shadow play told the story of a domestic worker leaving a rural tropical village in the Philippines, crossing the ocean by ship (represented by fish leaping across the waves), to reach a suburban Canadian home (Figure 1.10). A critical scholarly reaction informed by postcolonial theory might be this: without a tradition of shadow puppetry in the Philippines, its use is an orientalising, homogenising gesture across Asian cultures; and representations of the Philippines through the tropical rural (rather than urban Manila) reinscribe stereotypical notions of difference between the Philippines and Canada through binaries of rural/urban, and primitive/technologically advanced. However, when the PWC of BC held a community assessment following the play in March 2009, a long-term organiser at the Centre offered her reaction:

> In terms of content I also would like to express appreciation for the visual part of it. Especially the puppet show. Because I appreciated

Figure 1.10 Playing in shadow.

> having the chance to have different forms of expression within the one play. For me having heard a lot of the stories from the women it was also nice for me having it coming to me in a different form. That's also the part I got very emotional [...] the whole pictorial visual depiction of migration.

Two: we created one monologue in the play from an interview with a young woman who was separated from her mother for many years, while her mother worked in Vancouver as a live-in caregiver and she remained in the Philippines in the care of her father. Though verbatim, we had edited this monologue to tell a story of the problems experienced by Filipino youths in Canada, highlighting her urge to drop out of high school, her conflict with her mother and her compromised ambitions. Though she was enrolled in university in the Philippines, her ambitions shrank in Canada, and at the time of the interview she was pleased to have completed a six-month medical assistant course, selected in part because of its modest length and cost. The two youths from the Filipino-Canadian Youth Alliance, who had interviewed this young woman as part of the research on which the play was based, were unsettled when they first experienced the monologue. They noted that in the actual interview the young woman had been optimistic, expressing her personal triumph over what she saw as the destiny of most youths migrating to Vancouver to join their mothers who had come to Canada as migrant domestic workers. Through selective editing of her verbatim transcript, we had transformed her narrative into one of compromised success, indeed partial failure, and she was effectively rendered as a victim of the Philippines' labour export policy and Canada's LCP.

Three: we return to the Colombian woman's visceral reaction in the domestic worker's bedroom and her sense of profound violation as white audience members indulged their curiosity and the pleasures of looking through a domestic worker's private belongings and inhabiting her space at will. As Katherine McKittrick has argued in relation to the widely circulated image of the scarred back of a former slave, "if we are not very careful, the image becomes so ordinary that the pleasures of looking, again and again, incite a second order of violence."[56] How much more so if the audience member is invited to inhabit a domestic workers' private space and take momentary possession of her belongings.

We purposively offered no programme of action. Rather than a didactic event, we understood the play to be providing the materials – space, time, narratives, sensory experiences – for audience members to get an inkling of the tethered fates of employers and migrants and to have the kind of complex conversations that might move towards politics and to imagine open-ended futures that do not repeat the violence of temporary migration within racial capitalism and the ongoing devaluation of the racialised gendered labour of social reproduction. We aimed to bring Canadian audiences to an uncomfortably intimate recognition of the circumstances that make their childcare and eldercare solutions possible. And yet, we were (productively) left after the Vancouver and Berlin productions with many questions about the limits of our own imaginations, and the ways in which our own framings – theoretical and otherwise – may have unwittingly reinstated what we had hoped to dismantle.

Notes

1 Adrienne L. Burk, *Speaking for a Long Time: Public Space and Social Memory in Vancouver* (Vancouver: University of British Columbia Press, 2010).
2 Juan Manuel Sepulveda, *The Battle of Oppenheimer Park* (Mexico: Fragua Cine, 2016), Video recording.

3 This story has been the subject of two documentaries: Jari Osborne, dir. *Sleeping Tigers: The Asahi Baseball Story* (Canada: National Film Board, 2013) and Yuya Ishii, dir. *The Vancouver Asahi* (Japan: Pony Canyon, 2014).

4 One member of the Filipino-Canadian Youth Alliance described the steady traffic between the Kalayaan Centre and the theatre as we developed and rehearsed the play: "I see people going back and forth to the Chapel Arts, but I really never stopped by because I wanted to be surprised a bit. You know, 'Come, come, it's only a block across from the Centre.' And I say, 'Ah, I'll just wait so I don't ruin it for myself.'"

5 When the play was restaged at the HAU in Berlin, the same setup was approximated: the Canadian employers were in a relatively richly appointed public area of the theatre, the domestic worker scenes in behind-the-scenes working regions of the theatre. The only scene in which the actual theatre was used was a monologue by a nanny agent, not included in the Vancouver production. The CIC agent and youth scene and talkback were on the stage, literally behind the curtain.

6 There were three ways through the play: the audience of 50 was separated into three groups: 25 went upstairs to experience the employer monologues, and the remainder split into smaller groups of roughly 12 to go through the domestic worker scenes in one of the two orders. Those who saw the domestic worker scenes were then combined to see the testimonial scenes upstairs, and the larger group upstairs was divided into two smaller groups to witness the domestic worker scenes.

7 The monologue is developed from the interviews with Liberty, which is a pseudonym chosen by the interviewee during the original research. The character is renamed Ligaya in the play. For the entire script, see Geraldine Pratt and Caleb Johnston, in collaboration with the PWC of BC, "Nanay (Mother): A Testimonial Play," in *Once More, With Feeling: Five Affecting Plays*, ed. Erin Hurley (Toronto: University of Toronto Press, 2014), 49–90.

8 The two actors playing Filipina domestic workers had to perform their monologues four times in each performance because the audience is divided into small sub-audiences for their monologues. With four performances on the Saturday, this made for 16 performances altogether.

9 Alexander Ferguson, "Improvising the Document," *Canadian Theatre Review* 143 (2010): 35–41.

10 Inaccurate, as we found out later: this woman eventually was able to stay in Canada. In the Vancouver production, the guides were Filipino-Canadian activists who had been involved in the research as interviewees and focus group participants: Michelle Co, Charlene Sayo and Carlo Sayo.

11 Those enrolled in the LCP can change their employers and reapply for a new work permit from within Canada; this was a fought-for change to the programme. Still, having to complete the 24 months of registered care work within a defined period constrains the number of times that one can change jobs within the LCP (particularly because it can often take up to six months to obtain the next job permit). It is not unusual therefore for domestic workers to stay with exploitative employers. The time allowed to complete the required 24 months to qualify for consideration for permanent resident status was extended from 36 to 48 months after this interview was done.

12 For access to this audio recording, see Caleb Johnston and Geraldine Pratt, in collaboration with the Philippine Women Centre of British Columbia, "Nanay [Mother]: A Testimonial Play," *Cultural Geographies* 17, no. 1 (2010): 123–133. We composed this audio installation in collaboration with Vancouver scenographer Andreas Kahre.

13 The estrangement of children from their mothers because of the length of separation is a major focus of the play and the research upon which the play is based. The median number of years of family separation is eight, in part because many Filipina women go first to work as domestic workers in a country in Asia or the Middle East to gain sufficient capital (money, skills, contacts, information) to move up the migration hierarchy to Canada, and in part because it typically takes longer than 24 months to complete the LCP and much longer to process the papers to apply for permanent resident status and to sponsor dependents.

14 The mural collage was first assembled by the Sinag Bayan Cultural Arts Collective for their Maleta art exhibit at Gallery Gatchet in 2007 (www.straight.com/article-112540/lost-and-found).

15 Baucom provides an account of the *Zong* massacre, the landmark court case, and the political and cultural struggles that followed. A calculation of profit and risk led the captain of the *Zong* in 1781 to throw 132 slaves overboard, with the prospect of realising through an insurance claim their value was "jettisoned cargo." "To the extent the case of the *Zong* was to help define the struggle between slave traders and abolitionists in the late eighteenth century, the way in which that struggle was waged suggests that it was not only a struggle between competing theories of right (the slave's right to human dignity and the slavers' right to trade), but one between competing theories of knowledge, a struggle between an empirical and a contractual, an evidentiary and credible epistemology." See Ian Baucom, *Specters of the Atlantic: Finance Capital, Slavery, and the Philosophy of History* (Durham: Duke University Press, 2005), 16.

16 See Nicholas Abraham and Maria Torok, *The Shell and the Kernel: Renewals of Psychoanalysis*, ed. and trans. Nicholas T. Rand (Chicago: University of Chicago Press, 1994).

17 Ian Baucom, *Specters of the Atlantic: Finance Capital, Slavery, and the Philosophy of History* (Durham: Duke University Press, 2005), 132.

18 Ibid., 59.

19 One audience member describes her experience: "In our viewing, she actually poured a full bucket of water (for her mop) into the drain in the centre of the room, and it splashed up onto the trouser hems of a few audience members, and licked the shoes of everyone in the room."

20 Ian Baucom, *Specters of the Atlantic: Finance Capital, Slavery, and the Philosophy of History* (Durham: Duke University Press, 2005), 230.

21 Given the high educational qualifications of their mothers (most of whom have some post-secondary educations, often a university degree), the high drop-out rates of Filipino youths (boys in particular) has drawn scholarly attention. For instance, of the Tagalog-speaking boys enrolled in Vancouver high schools from 1995 to 2004, only 64% completed high school, one of the lowest rates of completion of any identified group. See Geraldine Pratt, *Families Apart: Migrant Mothers and the Conflicts of Labor and Love* (Minneapolis: University of Minnesota Press, 2012); Teresa Abada, Feng Hou, and Bali Ram, "Ethnic Differences in Educational Attainment among the Children of Canadian Immigrants," *Canadian Journal of Sociology* 34, no. 1 (2009): 1–29; Philip Kelly, "Understanding Intergenerational Social Mobility: Filipino Youth in Canada," *IRPP Study* 45 (2014): n.p.

22 See Martin F. Manalansan IV, "Queering the Chain of Care Paradigm," *Scholar and Feminist Online* 6, no. 3 (2008): n.p.; Ethel Tungohan, *From the Politics of Everyday Resistance to the Politics from Below: Migrant Care Worker Activism in Canada* (Urbana: University of Illinois Press, forthcoming); Ethel Tungohan, "Reconceptualizing Motherhood, Reconceptualizing Resistance," *International Feminist Journal of Politics* 15, no. 1 (2013): 39–57.

23 Linda Ben-Zvi, "Staging the Other Israel: The Documentary Theater of Nola Chilton," *Drama Review* 50, no. 3 (2006): 42–55.

24 In almost one-third of the questionnaires that were returned with comments, there was a question about the representativeness of the testimony. Some audience members were simply disgruntled because they felt that the play did not represent their experience: they had themselves been a good employer, had as a child loved their own nanny intensely, or simply imagined that some experiences were good. They could not imagine an epistemological framing in which we were not speaking about all Canadian employers, all domestic workers, or some average or typical experience. Our dramaturge anticipated the audience desire for typicality and balance and urged us to include a testimonial from a domestic worker who had had a positive experience. (This we did, but the director cut this testimonial because of the difficulty of accommodating it within the time constraints.) Some audience members asked for a positive example from the same desire to help us strengthen the persuasiveness of our case against the LCP by demonstrating our balanced perspective and hence objectivity and credibility: "Obviously the point of the show is to highlight the horrendous experiences of the LCP participants, but by not sharing any positive stories, you put the question 'What about those who aren't treated this way?' in the minds of the audience, which is distracting from the central message. Better to address it than leave even a friendly elephant in the room." From another, "It would be good to hear more about those who the program worked for/had okay employers who can also complicate their experiences. Not a romantic rosy view but a positive critical view. Otherwise I fear some could dismiss the very real experiences as hyperbole." In truth, this is difficult question or concern to answer because the LCP has been so poorly monitored or regulated. We know that there are Filipino activist organisations across Canada addressing issues of exploitation and abuse (See Ethel Tungohan, *From the Politics of Everyday Resistance to the Politics from Below: Migrant Care Worker Activism in Canada* (Urbana: University of Illinois Press, forthcoming). Regardless of whether abuse is the norm or exception (or something in between) what can be said with certainty is that the conditions of the programme create the circumstances in which this abuse is possible: i.e., little state regulation, a female employee living in their workplace and pressure to stay with an employer in order to complete the 24 months required to qualify for permanent resident status.

25 Boltanski analyses the necessity of this paradoxical slippage between singularity and the exemplary for politics. See Luc Boltanski, *Distant Suffering: Mortality, Media, and Politics* (Cambridge: Cambridge University Press, 1999).

26 Using the word *nanay* also captures the slipperiness of domestic workers' own passage between mother and servant. It should be noted, however, that the process of naming the play was by no means straightforward and was resolved through an uneasy and deadline-imposed compromise among collaborators. PWC of BC board members and the members of SIKLAB felt that *Nanay* was too "benign" and preferred the title *Atsay* (which they translate from Tagalog as "maid, slave, servant"). Their reading of *Nanay* as benign in part reflected their linguistic competence, which led them to interpret the word within its stable meaning in Tagalog and not for its mercurial slippage between English and Tagalog. A Tagalog word perhaps, but one that works in a particular way in the context of an English audience as a sobering example of the politics and limits of intercultural translation. The name of the play met with some resistance from English speakers as well. For a long time, the working title of the play was *Homekeeper*. When the suggestion was made to use a Tagalog word, some worried that this would turn away English-speaking audiences.

27 Another jotted on a questionnaire as he or she went through the show: "I am glad that there are members of the Filipino community in the audience. I'd like to hear their reactions."

28 See Alexander Lazaridis Ferguson, "Improvising the Document," *Canadian Theatre Review* 143 (2010): 35–41; "Authenticity and the 'Documentive' in Nanay: A Testimonial Play," *Platform: Journal of Theatre and Performing Arts* 11 (2017): 88–109.

29 Judith Butler, *Giving an Account of Oneself* (New York: Fordham University Press, 2005), 46.

30 When the actor who was meant to play this role in Manila was unable to do so, we approached the friend whose interview had been used to develop the monologue to see if she might allow us to record her telling her own story. She declined, citing the difficulty attending the performance in Vancouver and seeing and hearing her words performed as monologue. This raises an urgent issue of representation that is relevant to all social science research and sharpens the politics of representing the Canadian employers.

31 The staging in the Berlin production (more below) in our view more successfully invited audience members to empathise with a complex person. The actor delivered the monologues dressed in her own clothes, beside projected images of herself as a child in the company of her own mother and of her actual mother ageing over the course of time. (For a further discussion of the staging of this scene in Vancouver and Berlin, see Alexander Ferguson, "Improvising the Document," *Canadian Theatre Review* 143 (2010): 35–41).

32 Judith Butler, *Giving an Account of Oneself* (New York: Fordham University Press, 2005).

33 See Michael Anderson and Linden Wilkinson, "A Resurgence of Verbatim Theatre: Authenticity, Empathy, and Transformation," *Australasian Drama Studies* 50 (2007): 153–169.

34 See Jacques Rancière, *Disagreement: Politics and Philosophy*, trans. Julie Rose (Minneapolis: University of Minnesota Press, 1999); *The Politics of Aesthetics: The Distribution of the Sensible*, trans. Gabriel Rockhill (London: Continuum, 2004). Rancière is a natural ally for theorising *Nanay* because theatrical metaphor runs through his theorising of equality and emancipation, so much so that Peter Hallward applies to it the Platonic term "theatrocracy." But even more important, Rancière is alert to theatre, not just as metaphor but as a concrete space and practice, and as a site of politics. See Peter Hallward, "Staging Equality: On Rancière's Theatrocracy," *New Left Review* 37 (2006): 109–129.

35 This is central to Rancière's notion of egalitarian and democratic politics. Politics exist, he argues, when people who do not count or have a fixed place within the social order demand to be included in the public sphere, to be seen and heard on an equal footing. Politics is not the business-as-usual of contestation between already existing interest groups; it is a fundamental disruption of the existing "distribution of the sensible." With this phrase, the distribution of the sensible, Rancière is referring to the limits of what it is possible to see, hear and say within existing social arrangements. What he calls "politics" arises at the meeting of two logics: "egalitarian" and "police." He does not intend the latter term to be understood pejoratively, and it does not refer, in the first instance, to the legitimate violence or disciplinary practices of a state apparatus. The term "police" refers to social classification, the distribution of private and public space, occupational hierarchy and normative patterns of inclusion and exclusion. A social scientist investigating the settlement and integration of immigrants or a non-governmental organisation implementing multicultural policy are equally part of this police process of naming and sorting bodies into properties, tasks,

functions, occupations and places. As Rancière notes, some police are better than others and police have the capacity to "procure all sorts of goods" (Jacques Rancière, *Disagreement: Politics and Philosophy*, trans. Julie Rose (Minneapolis: University of Minnesota Press, 1999), 31). Although political subjectification is not, by Rancière's reckoning, anchored in specific identities or properties (indeed, the aim of politics is the disruption of classification schemes in the existing order), the structural conditions of migrant domestic workers make them likely prospects for politics because they are neither quite inside nor outside the nation. Migrants' articulation of their exclusion from and within an existing socio-political order is a performative contradiction: it performs their equality as speaking human beings with political capabilities at the same time as it speaks of their exclusion from the socio-political order. This articulation of political rights is different from claiming victim status; it is in itself an assertion of political capability and a demonstration of equality. For Rancière, equality is not an ontological principle that politics "presses into service" (Ibid., 33); it is discerned only in the practices that make it manifest. That is, it is a presupposition about the universal that is in need of constant verification.

36 The concern was unfounded. Almost all performances were at capacity. Over the 13 performances in Vancouver, 577 people saw the play. There were six performances in Berlin.

37 Quoted in Peter Hallward, "Staging Equality: On Rancière's Theatrocracy," *New Left Review* 37 (2006): 119.

38 Jestrovic discusses two fascinating performance events that trade on the cache of using actual refugees, asylum seekers and immigrants: a fashion show in Barcelona featuring illegal immigrants from West Africa as models, and Christoph Schlingensief's public art project *Foreigners Out!* Although at first blush both performances seem highly exploitative, Jestrovic makes the case for a more complicated ethical and political assessment. (Silvija Jestrovic, "Performing Like an Asylum Seeker: Paradoxes of Hyper-Authenticity," *Research in Drama Education* 13, no. 4 (2008): 159–170.)

39 We are referencing Rancière's definition of democracy: for him, the term designates neither a form of society nor government but the utter contingency – that is, equality – of those who govern and those who are governed. In other words, in democracy, government elites have no natural claim to their authority. The democratic public sphere is a sphere of encounter in which the logics of politics and police come into contest, in which the contingency of social relations is enacted. He argues that there is a tendency for any particular government to shrink the public sphere, "making it its own private affair," privatising and removing from common debate and consideration many of the concerns of ordinary citizens (Jacques Rancière, *The Hatred of Democracy*, trans. Steve Corcoran (London: Verso, 2006), 55). Democracy, in Rancière's view, is the struggle to enlarge the public sphere, to redistribute public and private, both by publicly performing the equality of those subjected to government and by establishing the public character of spaces, relations and institutions (such as the family) that have been relegated to the private sphere. "Democracy really means, in this sense, … the challenging of governments' claims to embody the sole principle of public life and in doing so to be able to circumscribe the understanding and extension of public life" (Ibid., 62).

40 In another talkback session, this feeling of being destabilised was expressed in this way: "As a theatregoer, to be led through the different rooms it felt like I was on a tour of a famous old house and there were actors playing the characters in the house, except it wasn't a famous old house, it was a house here in Canada and the actors were playing real people who were working in that house. So it was very … it's clear to me that it was a very clinical process of going through these different rooms and looking through materials and sort of being an observer

and sitting and watching as the theatregoer and patron. And then at the same time ... it's very personal and immediate to people. So it's an experience where you're wearing two hats. It's obviously something that's going to affect someone you know or someone you're met or perhaps you or someone in your family. But the process of seeing it as a performance is detached. ... It's interesting to try to understand how to read it as an audience member."

41 Gerry facilitated the talkback (she was already in the play as interlocutor of the child of the LCP) with a member of the Filipino activist community. In Vancouver, Carlo Sayo, chair of the Filipino-Canadian Youth Alliance, and Charlene Sayo, co-facilitated talkbacks. In Berlin, Dinah Estigoy, then of the PWC of BC, performed this function.

42 Jill Dolan, *Utopia in Performance: Finding Hope at the Theater* (Ann Arbor: University of Michigan Press, 2005), 2, 8, 21.

43 Quoted in Peter Hallward, "Staging Equality: On Rancière's Theatrocracy," *New Left Review* 37 (2006), 117; the embedded quotation is to Jacques Rancière, "Dix theses sur la politique," *Aux Bords du Politique*, 2nd edition (Paris: Editions Osiris, 1998), 242.

44 We thank everyone involved, including the Philippine embassy representatives – who could have felt under attack and easily removed themselves from the discussion – for their willingness to engage in such a vigorous public discussion in the talkback, which lasted until 11:00 pm (and into the bar afterwards).

45 Jacques Rancière, *Disagreement: Politics and Philosophy*, trans. Julie Rose (Minneapolis: University of Minnesota Press, 1999); 55, 56.

46 Dean's argument is two-pronged. She argues that Rancière's political theory is both sociologically naïve and political misguided (the latter for the reason already mentioned). It is sociologically naïve in her view because it assumes some level of mutual understanding that is "rather far-fetched." She argues that "our present political-medialogical setting is one of dissensus, incredulity, and competing concepts of reality." In making this claim, Dean draws on David Donaldson's "principle of charity." This principle posits as a necessary background assumption that, in order to understand another, we have to "count them right in most matters." Dean argues that this principle of charity has "withered away" and for good "material-technological" reasons: so much of our communication is now mediated by machines that are "charitable in our stead," and these machines allow us to create and restrict our communication to those in like-minded communities. We increasingly communicate, often via electronic media, within geographically dispersed but narrowly conceived, like-minded communities. Fewer and fewer common assumptions are hold, she contends, across these different worlds. In arguing this, Dean actually helps make the case for the significant time-space presence of a theatrical experience such as *Nanay* and the importance of the affective intensity of witnessing across social worlds. See Jodi Dean, "Politics Without Politics," *Parallax* 15, no. 3 (2009): 32.

47 Peter Hallward, "Staging Equality: On Rancière's Theatrocracy," *New Left Review* 37 (2006): 125, 128. For other useful engagements with Rancière's theorising on aesthetics and politics, see Timon Beyes, "Uncontained: The Art and Politics of Reconfiguring Urban Space," *Culture and Organization* 16, no. 3 (2010): 229–246; Claire Bishop, *Artificial Hells: Participatory Art and the Politics of Spectatorship* (London: Verso, 2012); Mustafa Dikec, "Beginners and Equals: Political Subjectivity in Arendt and Rancière," *Transactions of the Institute of British Geographers* 38, no. 1 (2013): 78–90; Alan Ingram, "Rethinking Art and Geopolitics through Aesthetics: Artist Responses to the Iraq war," *Transactions of the Institute of British Geographers* 41, no. 1 (2016): 1–13; Naomi Millner, "Activist Pedagogies through Ranciere's Aesthetic Lens," in *Geographical Aesthetics Imagining Space, Staging Encounters*, eds. Harriet Hawkins and Elizabeth Straughan (London: Routledge, 2015), 71–90; Divya P. Tolia-Kelly, "Rancière and the

Re-distribution of the Sensible: The Artist Rosanna Raymond, Dissensus and Postcolonial Sensibilities within the Spaces of the Museum," *Progress in Human Geography* (2017): early view.

48 Jill Dolan, *Utopia in Performance: Finding Hope at the Theater* (Ann Arbor: University of Michigan Press, 2005), 19–20, emphasis in original; see also Erika Fischer-Lichte, *The Transformative Power of Performance: A New Aesthetics* (New York: Routledge, 2008).

49 Tom Burvill, "'Politics Begins as Ethics': Levinasian Ethics and Australian Performance Concerning Refugees," *Research in Drama Education* 13, no. 2 (2008): 234.

50 Lauren Berlant, "Thinking about Feeling Historical," *Emotion, Space and Society* 1, no. 1 (2008): 5, 6.

51 Rancière quoted in Peter Hallward, "Staging Equality: On Rancière's Theatrocracy," *New Left Review* 37 (2006): 115.

52 Other comments included the following: "I used to run a Home Depot with 100–300 employees being Philippino. They were a very tight community. This helps me to gain insight into their community"; "As someone who has worked in the service industry for many years, I worked with many philipinas. We always valued their hard work and ability to hold down several jobs concurrently. I remember hearing stories about working for families or hearing about caring for elderly people, but I never gained such an insider view from the stories. Many of the comments that I heard at work were somewhat fatalistic and along the lines of – 'Oh, I just need to do this' and 'This is how it works.' I wish more of my co-workers could have shared their *real* feelings about how they were treated"; "I have a better understanding now of why so many Philipino boys in our high schools are so detached and have so much trouble succeeding"; "I have elderly parents requiring significant care and a sibling is pushing for an LCP caregiver. This play provided a bit more personal detail to support principled argument against my sibling, which may help convince her."

53 Geoff Meggs, "Nanay: Filipino Word Meaning 'Mother,'" *Geoff Meggs: Vancouver City Councilor* (blog), 9 February 2009.

54 To contextualise this statement: "While the physical space and the location of the different stories emphasise the two solitudes between the employers and the live-in caregivers, I also felt that the play was a very good opportunity for us to try to bridge those two groups. My personal experience of course has been as an advocate in the [Filipino] community. But I also grew up here in Canada. I work in different provincial governments. So a lot of the people I actually come into contact with are employers. But the play gave me more of a space to try to understand how we can bridge those two groups while appreciating the distinction between the classes, between races, between genders. I think that the talkback really emphasised for me that it's a human rights issue and it's a woman's issue. So as a mother who is a working mother, as a Filipina, it was a really good chance for me to see how those different perspectives can perhaps come together in the long term: that we have child care that is accessible for all working women here in Canada."

55 See Hartman and McKittrick for the ways that the analytics of violence and anti-blackness can insinuate themselves within narrative attempts to reveal and redress violence. (Saidiya Hartman, "Venus in Two Acts," *Small Axe* 26 (2008): 1–14; Katherine McKittrick, "Mathematics Black Life," *The Black Scholar: Journal of Black Studies and Research* 44, no. 2 (2014): 16–28).

56 Katherine McKittrick, "Mathematics Black Life," *The Black Scholar: Journal of Black Studies and Research* 44, no. 2 (2014): 21.

2 Travelling with baggage to the Philippines

After the presentation of *Nanay* at Vancouver's 2009 PuSh International Performing Arts Festival, we met with the director and migrant organisers from the PWC of BC to assess the project and our collaboration. For many domestic workers, watching audiences in Vancouver had prompted the desire to tell their stories of life in Canada differently to those they care most about: their families whom they had left behind in the Philippines. There are many reasons why the experiences of migrants may not be told or known within their transnational families: shame that a nursing degree, for instance, has led to a job that includes cleaning toilets in a private home, a desire to protect family members from worry and the pain of family separation, or a personal investment in maintaining the fiction that Canada is the unqualified "greener pasture" or "dream destination" that it is typically assumed to be in the Philippines. Further, families that rely on remittances to ensure their day-to-day survival may not want to fully know the details of what is required of their loved ones who labour overseas. We hoped that taking the play to the Philippines would be one way for domestic workers at the PWC of BC to fill silences within their families about their lives as domestic workers in Canada, and to circulate the stories – often painful stories – that can go unheard in the Philippines.

Taking *Nanay* to Manila brings a more nuanced, troubled – and we think accurate – impression of Canada to the Philippines. Because of its proximity to the United States – literally and culturally, and the opportunity the LCP (now Caregiver Program) affords to migrate permanently, Canada is a top destination within a hierarchy of migration destinations and migrants must activate considerable resources to get there.[1] While the opportunity to attain permanent residency (and ultimately citizenship) in Canada through the LCP is by no means trivial, the opportunities for exploitation and abuse within it are no less so, and communicating this is important, both (potentially) for the Philippine state's efforts to regulate the conditions under which its citizens labour as OFWs, and to migrant organisations in the Philippines critical of the Philippine state's labour export policy. *Nanay* provides evidence that even one of the most coveted migration destinations has serious problems of systemic racism and labour exploitation. Our insistence on telling this story in the Philippines was further propelled by the reactions of representatives of the Philippine government to an earlier production of *Nanay*. In a talkback following one performance, they had affirmed Canada's place at the top of the migration hierarchy and downplayed the problems raised in the play. This seeming refusal ignited our commitment to travel to the Philippines with the play.

Closer to our institutional homes as academics, travelling with these stories to the Philippines speaks to and works against what have been identified as failures within Asian American, migration and transnational studies that follow from an exclusive focus on the so-called receiving country. Oscar Campomanes has criticised the "contributionism" and "domestication" of Asian American studies.[2] Andreas Wimmer and Nina Glick Schiller are critical of the "methodological nationalism" that characterises much of migration studies.[3] Within postcolonial or feminist transnational scholarship, Gayatri Spivak writes of the "narcissistic, question-begging" reductions of migrant lives to the categories of race-gender-class, that is, to the structured ideological framings of the global north.[4] These criticisms highlight the parochialism of much of Anglo-American scholarly work and its simplification of migrants' positionings, predicaments and attachments. Focusing only on migrants' lives in North America can effectively contribute to the self-valorisation of the United States or Canada as nations of immigrants and liberal multiculturalism. It has had the effect, Campomanes argues, of "defanging" Asian American studies because it displaces attention from the implication of the United States in the histories of colonialism, imperialism and capitalist uneven development that drive migration. The original script of *Nanay* was vulnerable to these criticisms, and taking the play to the Philippines was an opportunity to broaden the frame of analysis.

We had intimations that the play might not, however, travel easily and that we would bring more baggage than intended or desired. Seen in Canada, the play brought into presence an issue that many can and do comfortably ignore, namely, Canada's reliance on a type of indentured servitude to address national childcare and eldercare needs. In the Philippines, most everyone possesses an intimate personal connection to migration. Rather than a lack of feeling and visibility, migration is hypervisible in the Philippines, and the story is, if ever changing, well worn, even hackneyed. Overseas migration, Filomeno Aguilar writes, is "like yeast that has gradually worked its way through Philippines society, turning things around, throwing up established practices, revising assumptions about sociality."[5] The success of the Philippines as a labour brokerage state relies on its capacity "to standardize and script the Filipino as a sought after commodified figure"[6] and much state and media discourse has been expended, domestically and internationally, on these scripts and positionings. Possibly because the migration of so many women as OFWs has caused national anxiety and controversy, Filipino domestic workers in particular, Lorente argues, are likely "the most extensively 'branded' and [...] scripted Filipino bodies circulating in the global marketplace."[7] Meeting with a colleague at the University of the Philippines (UP) Diliman in 2012, she spoke of a generalised fatigue within the Philippines about the ways that migration stories tend to be narrated in both popular and academic discourse: migrants tend to be portrayed as either victims of abuse and exploitation or heroes and saviours of the nation.[8] Eager to rework the script to move towards more nuanced transnational conversations, we made preparations to make *Nanay* travel-worthy, changes that we detail below.

Even so, although the play was created in collaboration with a Filipino activist organisation in Vancouver to address the experiences of Filipino domestic workers and their children in Canada, and we brought the play with Alex Ferguson to Manila as Canadians (and not as interpreters of Philippine society or culture), the play's travels bear all the trappings of our privilege. In the first instance, we had access to the resources to fund its transnational itinerary.[9] Writing about the uncomfortable ambivalences and dilemmas arising when he brought students (a good number of whom were Filipino Americans who understood the Philippines to be their homeland) from the University of Washington to the Philippines in 2007 and 2011, Rick Bonus notes that he insisted that his students recognise the ongoingness of U.S. colonisation in the Philippines as "precisely the condition that ma[de] possible all aspects of their study-abroad trip."[10] He came to understand the course he taught as a "contact zone" of intermingling cultures in which the students' privilege could never be forgotten, but which they persistently and sometimes unwittingly re-enacted. Taking *Nanay* to the Philippines offered many opportunities of learning and unlearning our privilege as scholars and theatre practitioners from the global north. In Vancouver, we had created *Nanay* as an intercultural space in which to bring Canadian audiences (including employers) closer to the felt experiences of migrant domestic workers (and vice versa) and to stage public debate on the issues. Travelling across the Pacific with the project moved us into a transnational frame, and it was the physicality and materiality of the play – the concrete bringing together of bodies, translators and audiences, activists, government representatives, migrants and family members – that highlight the possibilities and limits of transnational analysis and of border-crossing solidarities.

Contemplating Derrida's text, *Postcards*, Sara Ahmed notes that the "potentiality" of non-arrival is ever-present: "If you take from this idea that the purpose of communication is not always an arrival, the point is not always to reach something, which is already known in advance of sending something out, it changes the meaning and the conduct of the conversational space. One is no longer confident about what it means to succeed in an act of communication... That is precisely the point of translation [as well]. Because in a way it's about the creation of a new text."[11] Despite our efforts to revise the script and staging for Manila, our efforts at translation were partial and inevitably flawed. If translation is in some sense always incomplete, always impossible, taking and resituating *Nanay* within the Philippines was also generative; it opened up the possibility for the unexpected, not an arrival point or destination but for new conversations in a multidirectional conversational space with Manila-based theatre artists, migrant advocacy groups and audience members, including the family members of domestic workers in Canada. We turn to our preparations for these conversations.

Relocating to PETA Theater Center

In fall 2013, we began a collaboration with PETA to stage *Nanay* at their theatre in Quezon City in Metro Manila. The director, stage manager and three actors came from Vancouver (the actors and director to perform four Canadian employers, a Canadian government official and a nanny agent). A Filipino set designer was enlisted to help with the restaging, and three PETA actors and another Filipino actor (from Mindanao, associated with the National Commission for Culture and the Arts, and with a long history in community theatre) were recruited to perform the Filipino characters; these included two domestic workers, a child of a domestic worker who had migrated to Canada, and a representative of the Philippine government. PETA provided technical support, community outreach and space.

We were privileged to work with PETA. Since its founding in 1967, PETA has played a key role shaping political theatre in the Philippines through its plays, performances at political rallies, co-organising national festivals of peoples' culture, training, educational workshops and community advocacy. PETA was a leading force in the peoples' movement that eventually deposed President Marcos' dictatorship in 1986. Writing in the aftermath of that mass popular uprising, Eugene van Everen describes PETA as "the eloquent consciousness and pensive jester of Philippine society and the undisputed leader of its powerful contemporary theatre."[12] PETA is no stranger to local, national and international collaboration: as early as 1971 it received funds from UNESCO to expand the Central Institute of Theatre Arts in the Philippines throughout Southeast Asia, and as of 2001 the Canadian Catholic Organization for Development and Peace funded its work in schools, and the German Evangelical Mission and Terre des Hommes provided funds for their women's programmes.[13]

The PETA theatre also offered exciting dramaturgical and staging opportunities, and our collective work began with a two-week rehearsal process in which we focused on resituating the project within a new social, political and emotional terrain.[14] Reworking the Vancouver production, we redesigned the processional architecture of the play, plotting twisting routes and installing testimonial monologues through the back rooms, passage ways, rehearsal and office rooms scattered throughout the three-storied PETA complex. The opening of the show was significantly reconfigured. In Vancouver, audience members had gathered in the foyer of the Chapel Arts Centre where computer monitors and video projectors had been set up for spectators to peruse the websites of various Canadian nanny agencies and to scroll through the resumes of Filipino women who could be contracted and brought to Canada as domestic workers. In Manila, audiences met in the downstairs PETA lobby to view a projected video, which we created as a montage of clichéd Canadian scriptings and iconography (Figure 2.1). This included "fun facts" about Canada; an edited excerpt from a Government of Canada promotional film celebrating the country's cultural diversity; iconic panoramas of snow-topped Rocky Mountains and nightlife in Vancouver; and an assortment of "I Am Canadian"

Figure 2.1 Views of Canada.

commercials (a playful and ironic series of popular TV beer advertisements playing on a range of popular Canadian stereotypes: infatuations with ice hockey, long underwear for long winter months, rabid beavers, celebrations of diversity (not assimilation), peacekeeping (not policing) and so on). The intent of the pre-performance exhibition was to invoke entrenched stereotypes and preconceived notions of Canada, some of which we hoped to disrupt during the performance. As was the case with the Vancouver and Berlin productions, each audience member in Manila was given a survey to fill out as they moved through the testimonial archive. In order to glean spectators' relationships and insights to the issues in the Philippines, our revised survey questions included "I have employed a domestic worker"; "I have friends or family in Canada who have gone through the LCP"; or "I am interested in immigrating to Canada." The questionnaire provided opportunity for audience members to offer thoughts and criticism of the play. Members of PETA worked as tour guides, guiding groups of 20 or so through a maze of rooms; they wore white golf tee-shirts embossed with a bright red Canadian flag.

The director, Alex Ferguson, attempted to bring some of the intimacy and possibility of our transnational collaboration into the revised play and to the audience. For instance, he worked with the actor, Joanna Lerio, who performed one of the domestic worker scenes, to support her to feel more at home within the monologue and to bring it to life for a Filipino audience. The actor did this by translating portions of the monologue into Tagalog and introducing moments of audience engagement (Figure 2.2). When she tells with indignation, for example, that her Canadian employer insisted that she help her children with their French homework, she momentarily abandoned the monologue to teach an audience member some French words in a terrible French accent. The scene was performed in a stuffy backstage room representing a "safe" house where the domestic worker

Figure 2.2 Joanne in hiding.

was in hiding to avoid detection from the Canadian authorities (which was true to the circumstances in which she was interviewed). In another scene, the director merged what were previously separate domestic worker and employer scenes, bringing Canadian and Filipino actors and monologues together in a new way. The domestic worker tells of her great optimism leaving the Philippines to work overseas, tied to the opportunity that this creates to establish her own financial independence (from both her irresponsible and unfaithful husband, and her patronising father and brothers) and to earn an income that will allow her to send her children to good private schools in Manila. In the Vancouver and Berlin productions, this monologue was performed by a Filipina-Canadian actor working alone in close proximity to the audience. At PETA Theater, the director intermingled this monologue (performed by Marichu Belarmino in the backstage kitchen area of the theatre) with another delivered by a Canadian woman (performed by Hazel Venzon) who laments her history of trying to get dependable childcare from within the existing supply of Canadian labour, that is, without relying on Filipino live-in migrant domestic workers. This Canadian mother tells of how the unreliability of childcare provided by Canadian workers eventually led her to quit her job to care for her children at home. The two professional actors delivered their interspersed monologues in the same cramped space, at points mirroring each other's bodily movements in a slow-motion choreography. Their monologues essentially were kept separate, but there were occasional moments of surprised or curious visual recognition, intimating the possibility (and need) for a conversation about social reproduction and women's issues across global south and north (Figure 2.3).

When we began discussions about *Nanay* in Manila, we assumed that the employer scenes would be of less interest to audiences. But those we spoke with on our first trip to Manila in 2012 quickly disabused us of this notion, pointing out that what is less known in the Philippines is the perspective of Canadians. Further, if our goal was to disrupt the notion that Canada is an unproblematic multicultural nation, it was important to include the perspectives of employers and Canadian nanny agents. And so, we included a monologue of a nanny agent, not staged in the Vancouver show. Carl's testimony was staged in an upstairs office, complete with desk and computer. On the walls, we posted the head shots of European nannies which the actor (played by director Alex Ferguson) gestured to during the scene to highlight the differences between European and Filipino nannies. We printed business cards, which Carl (departing from the verbatim script, and breaking the boundary between actor and spectator) handed out to audience members as they departed the room, impertinently asking them to get in touch "if they were looking for work." This monologue is particularly difficult to experience because the agent is explicitly and brutally racist in the distinctions that he makes between European and Filipino domestic workers. For instance, "You ask other people; they'll tell you the same thing about

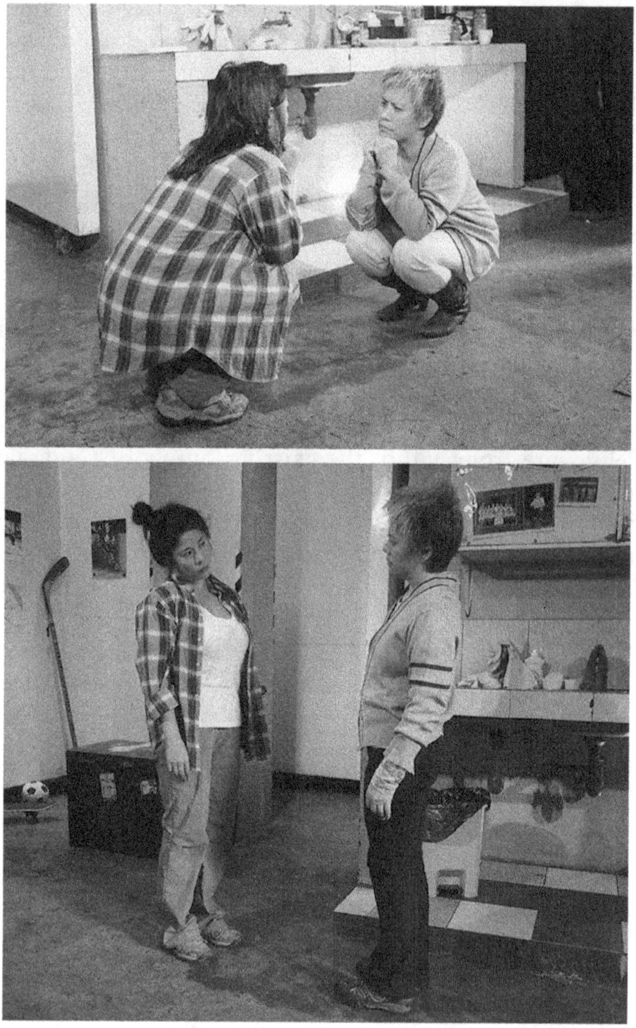

Figure 2.3 Mirrored movements.

Filipinos. Depends what you're looking for, what you want. My personal view, if you have a baby and you want someone to lick your home clean: Filipino girl. Go for that. If you have kids 3, 4 years of age, and you want interaction, you want them to go to the park, arts and crafts, do things, you're better off with a European" (Figure 2.4).

Presenting this monologue in the Philippines effectively forces Filipinos to listen to racism directed towards them. Reflecting our discomfort with the scene and possibly distancing ourselves and our responsibility for it, we had the guides who led audiences from scene to scene explicitly remind them

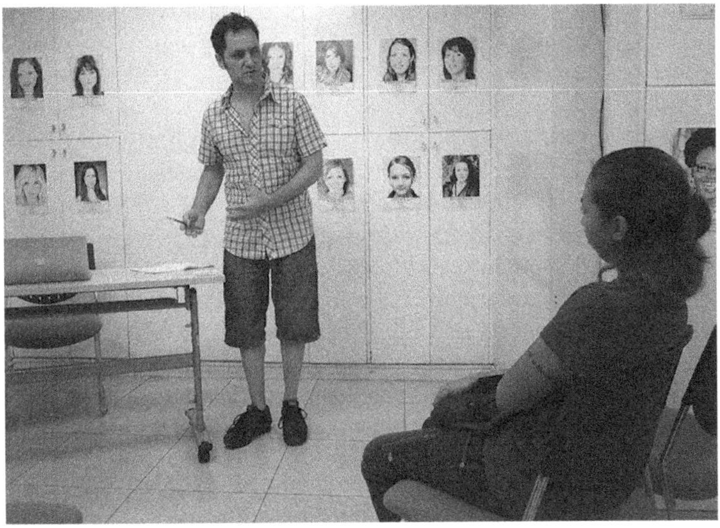

Figure 2.4 Performing racism.

in advance of entering the scene that this was a testimonial monologue – that is, these words were spoken by a real nanny agent in Canada and not a product of our own writing. All the same, it was performed by the actor in uncomfortably close physical proximity and the performed character engaged them not as middle-class audience members, but as potential migrants to Canada. The audience was awkwardly suspended between the theatre and everyday life. While we were uneasy about subjecting Filipino audiences so directly to Canadian-based hate speech, on balance, we decided that this shockingly overt racism brings the issue of systemic racism in Canada into the public conversation in difficult but important ways. It disrupts naïve fantasies of Canada so as to open room for – and even force – a more probing discussion of the Philippine labour export policy and its associated culture of out-migration.

To open conversations about Philippine state policy, we also built into the script verbatim comments made by Philippine government representatives who participated in previous post-performance public talkbacks of the play. In previous productions of *Nanay*, only the Canadian government "spoke," through the one constructed or non-verbatim monologue in the script. For the PETA production, we developed a chummy exchange between a Canadian and a Philippine government representative, the latter's lines developed verbatim from a talkback transcript from a previous production. This was performed by Patrick Keating and Lex Marcos, respectively, and staged in the Lino Brocka Hall, a multi-function connective space located at the intersection of the building. During this scene, multiple still photographs of a domestic worker appear in a PowerPoint presentation, photos of the actual mother of one of the family members who was able to come to Manila to view the performance. The script for this scene follows.

CITIZENSHIP AND IMMIGRATION CANADA (CIC) REP: [...] *The World Bank and many academic scholars now believe that the money sent home as remittances is a more effective means of stimulating the Philippines economy than more traditional forms of development aid.*[15]

PHILIPPINE FOREIGN AFFAIRS REP (PR): If I may, I would like to clarify something. It has never been the policy of the Philippine government to send domestic workers abroad. It is perhaps a consequence of the fact that they would like to look for so-called greener pastures. What we have always been trying to do is to keep people home where they belong. Now if they choose to go out of the country to work we need to cope with this. So, institutions have been put in place to protect the interests of migrant workers. Of course, the Philippines doesn't have all the resources to cover everything and attend to all the needs of these people. But if we compare the Philippines' performance with its neighbors, we are doing a lot more. In migration circles, our government is a model for other labor-sending countries. But it is not the policy to send domestic workers abroad. It is a personal choice to leave the country.

CIC: *Exactly! About 20,000 Filipinos are currently registered in the Live-in Caregiver program and lots more would come if they could. Shouldn't individual Filipinos have the opportunity to decide for themselves whether this is a good or bad program?*

GUIDE: May I ask a question? If one must migrate out of economic necessity, is it really a choice?

PR: If I may add, it just shows that people are free to leave if they want to leave. The official migration policy started roughly in the 70s. It was the brain-child of the late secretary Blas Ople. Before that we never heard of Filipinos going abroad en masse to work. It was just a temporary measure at the time when there was an oil crisis. Somehow it seemed to gain some roots.

But for the past nine years, it has not been the official policy of the Philippines to send labour abroad. Because the social costs far outweigh the remittances. We know that the separation of parents from their children and spouses from each other brings up drug abuse, early teenage pregnancy and other social problems. These problems far outweigh the monetary benefits that we get from the remittances abroad.

I would also like to add that conditions in Canada appear to be good compared to those in other countries. Listening to the stories in this play, you can really feel for these women, but they are in a programme that after a few years they can bring their families and they can immigrate.

If they work in other countries – in Korea, Taiwan and Japan – they never have this opportunity to be reunited with their families one day. So, they are really better off.

CIC: *Agreed! LCP is not perfect but it is the most generous program of its sort in the entire world. Canada is the only country in the world to offer citizenship after a short period of time.*

GUIDE: Again, pardon me. But am I correct in my understanding that the Canadian government has created another program for temporary foreign workers to come to Canada as fast food workers? And that these workers cannot stay.

PR: The Philippine government is trying to secure the best working conditions for all of its workers abroad. Before we criticize the new Canadian program for temporary foreign workers, we need to think carefully about the benefits and costs of permanent migration to Canada. We educate our people using Philippine government money. For what? For them to go abroad? That is a loss of capital even if you count all of the remittances coming in. So, from the point of view of the state we spend money to raise children to make them doctors. And then they are deskilled. They apply as nurses and become nannies. They will be doing a very good job because they are highly qualified – over qualified – in Canada. But it will be a loss to the Philippines. We have a very low doctor to patient ratio. How many percent of the population can see a doctor before they die? This is a massive loss of capital.

But these are things we are already working towards solving. We are trying to keep education as good and as strong as it has always been in this country and we are encouraging people to invest back in the country and in fact come home. It is unfortunate that countries like ours still have nationals who leave to work abroad. But our government is working towards making their protection the most important thing in their agenda.

As for the problems raised in this play about the LCP, this seems to be a Canadian problem.

Finally, we also attempted to address a cluster of concerns around the scripting of Filipina domestic workers as victims. We did this in the first instance by developing a monologue (also verbatim from an interview transcript) that tells a more positive migration story. A self-scripted go-getter, this woman extolled the virtues of her entrepreneurial individualism coming through the LCP to Canada while simultaneously narrating how she managed her homesickness and separation from her young children. She spoke of turning on her computer and using its webcam to listen to her children: "When I get homesick I just listen to the little noises they make, what conversations they are having, when the kids are playing. Even if the camera is not focused on them, I could still hear what they are doing around the house... That's all. I just listen." Her story of individual success is also held against community suffering:

JOVY: [Do] you know how many people from the Philippines get disappointed when they get here? [...] That's why so many of us caregivers who are first timers going abroad, [end up] going back [to the Philippines]. We learn about those cases. There are *a lot*. Either they commit suicide here from being so depressed or go crazy [or] return after eight months. [Or] it's almost at the end of their 24-month term and [they] couldn't make it and so went back. It's like, why? Almost 24 months and you quit? Types like that. If you're not strong here [points to head] and you're not strong here [points to heart], you'll have a hard time.[16]

Beyond balancing this individual success story against community hardship, we attempted to displace a victim narrative by showing some of the extraordinary organising by and vibrant sociality among domestic workers in Vancouver.[17] Migrante BC organisers allowed us to video record[18] a group meeting at Migrante House in Vancouver, which included laughing and sharing stories across a range of themes: first impressions of Vancouver, reminiscing about singing "Please Release Me" when they got together on weekends as domestic workers (they did a rousing rendition for the video) and their work as organisers in Canada.[19] In Manila, the director used some of this video material in the scene of a domestic worker's child to accentuate and make more complex the challenges of family separation and reunification. As Michelle (the daughter, played by Filipino actor Anj Heruela) tells of her loneliness in Canada and feelings of estrangement from her mother, she turns the sound of the video on and off, and addresses some of her monologue directly to the video image of the most prominent and expressive of these women, who is evidently enjoying the companionship of her sister domestic workers as she speaks about her life in Vancouver (Figure 2.5).

And lastly, to disrupt an easy rendering of Filipinos as pure victims, a Filipino-Canadian actor, Hazel Venzon, who previously had been cast as a domestic worker in Vancouver, was recast as a Canadian employer in the PETA production. The recasting of a Filipino-Canadian actor in the role as employer is an important and plausible scenario for at least two reasons. First, the actual interview upon which the monologue was based was conducted with an Asian-Canadian woman, and second, migrant organisers in Vancouver estimate that roughly 40% of the applications through the LCP in 2013 were being made by Filipino-Canadian families.[20] Casting a Filipino-Canadian actor as an employer complicates the racialisation of employers and employees in Canada and opens up for discussion

Figure 2.5 Feelings of estrangement.

the complex class dynamics within many transnational Filipino families. It disrupts one characteristic of "methodological nationalism": this is the tendency to disregard social and cultural divisions within a nation-state and to prioritise ethnicity over all other forms of identification through a so-called "ethnic lens."[21]

The staging of this scene was also reworked in Manila. At the Chapel Arts Centre in Vancouver, the director had situated the actors in a bedroom, dressing into formal evening attire while narrating paternalistic attitudes about their Filipino nanny: e.g., "we heard they were very caring for the very young ones." A tantalising (and concerning) opportunity to parody and ridicule the experiences of Canadian employers, the director had the actors drinking martinis, slowly becoming more and more inebriated as scenic images of downtown Vancouver were projected on a facing scrim. Their indulgence was accompanied by a pleasing melody of a breezy Brazilian bossa nova. In Manila, the employers remained in a bedroom. A plush circular bed fitted with red velvet sheets was placed in the middle of an upstairs PETA rehearsal studio. Several large square mirrors were positioned to form an octagonal pattern encircling the actors, who delivered the monologue while undressing to their underpants, and at intervals coming together to enjoy an intimate slow waltz. The action ended with actors laying down on the bed, becoming inanimate objects, like a still life in a museum or gallery, possibly both a reiteration and reversal of colonialist ethnological displays. Filipino audiences crowded into this small space and were encouraged to view and inspect the "exhibition" of Canadian employers (which they did with some confusion and discomfort) (Figure 2.6).

Figure 2.6 Employers on display.

Inviting the migration bureaucracy: a preliminary tour of white privilege

As Caleb worked at the PETA theatre, May Farrales (then a PhD student at UBC, and a research assistant on this project in Manila) and I (Gerry) worked on getting an audience out to the show by sending emails and texts, and hand delivering invitations. Visiting with Migrante International one day, organisers there suggested that we deliver invitations along a route they take migrants on to educate them about the migration bureaucracy. They gave us a detailed map of how to move from one office to the next, and we followed it the next day.

We began at the Department of Labor and Employment (DOLE), which oversees the Administration of the Philippine Overseas Employment Agency, to deliver an invitation to the Secretary of DOLE, Rosalina Dimapilis-Baldoz. We then moved on to the Department of Foreign Affairs (DFA) to deliver an invitation to Jesus I. Yabes, the Undersecretary for Migrant Workers' Affairs (OUMWA).[22] At both of these two stops, our bags were searched, and we presented passports and were signed in and signed out. Perhaps the most extraordinary thing is that we were admitted into the buildings, allowed onto elevators and to come close to our intended destinations. In the case of the OUMWA, Mr. Yabes was overseas, but we were greeted by a friendly and chatty young woman, recently appointed to administer visas in the Vancouver office of the Philippine Consulate. She is the third generation in her family to work for the DFA, her grandfather had been Consul in Vancouver years before, and she has many family members there. Business cards were exchanged.

As we left the second building, it began to rain and a well-dressed young man came alongside and offered to shelter May and I under his umbrella. When he realised that we were walking to the perimeter of the grounds of the DFA to hail a cab, he returned us to the building, bringing us through a door that requires no security. He instructed the receptionist to call us a cab and to tell the cab driver to come into the grounds to pick us up directly in front of the building. He too had links to Canada, having worked at the Canadian embassy in Manila. He had recently passed his Foreign Service Exam, one of a small cadre of successful examinees (less than 10% of all who write the exams).

The pleasures of being interpellated through elite transnational connections ended there, as we headed to spaces more frequented by migrants: the office of the Overseas Workers Welfare Administration (OWWA) and then the Commission of Filipinos Overseas (CFO), the latter where migrants attend pre-departure orientation seminars. The courtesy extended to us diminished: no more joking, exchanges of business cards or helpful umbrellas. Sitting on the ledge just outside the front door of the CFO, we met a 17-year-old youth holding a brochure about going to Canada. He was leaving in four days to join his mother in Vancouver. We asked whether he knew which high school he would be going to (with the challenge of starting halfway through the school year). His uncle (who was accompanying him) corrected us to say that he was already in university. May gently forewarned the youth that he likely would be enrolled in high school in Vancouver, but we left dispirited by the flows of misinformation, anticipating the all too familiar adjustments that lay ahead for this youth and his mother in Vancouver.

Stopping last at the Canadian Embassy, we came closer to and were more immediately subjected to the humiliations of the migration bureaucracy. The Canadian embassy is located in an office tower in Makati, and its first point of access is at a raised counter behind which sit male receptionists/security guards serving the embassies of different countries, including Canada, Australia and

Germany. When we reached the front of the line for Canada, at the counter we were faced with a list of fees printed on a yellow sheet of paper held upright at eye level in a hard-plastic frame. The list culminates with a fee of 1,500 pesos for "removal." We had had many fruitless email communications with the Canadian embassy in previous months in which we had informed them of and invited them to the play. Today, we said that we wanted to hand deliver an invitation to our play, a Canadian cultural production (partially funded by the Canadian government) to be staged at PETA Theater Center. The receptionist directed us to the drop-off area in the basement. Once in the elevator, it became apparent that the basement is five floors deep, which the receptionist had not thought to inform us. A fellow passenger on the elevator thought that the Canadian embassy might be on the third floor down. The doors opened and we entered an expansive underground parking lot. The guard on this level told us that the Canadian embassy is on the fourth floor. And so, we went further down. The elevator door opened and we were again in an underground parking lot. The guard on this floor directed us across the expanse of the parkade to an enclosed glass booth, with a Canadian flag drooping beside it. The shades inside the glass booth were drawn down. We walked around and around the booth. Thinking that we must be in the wrong place, we returned to the security guard, who redirected us to the booth. As we looked closer, we saw three labelled mail slots about thigh high in one side – one slot labelled LCP visa, another foreign worker visa and a third for other categories of visa. It dawned on us that we were supposed to stick the invitation in one of these slots. We then noticed a small opening at counter height, a rotating plate like one might find at a bank or money exchange, where you could pass a thin package or envelope to a willing recipient. As we stood there, the slats in the blinds opened slightly at about waist height. I crouched to engage the person in conversation. The slats snapped closed. We could hear a voice but could not discern what was being said. I shouted indignantly that I could not hear him. There was no response.

And there was no further communication. My indignation at the facelessness and the dehumanisation of the situation is perhaps the most interesting and possibly naive part of the story. A young Filipino man who had been before us in line upstairs, and who was applying for a temporary visa for construction work in Canada, joined us in the basement. May had overheard him saying to a friend in Tagalog after he spoke to the receptionist upstairs: "That was a waste of time." It sure felt that way.

Transnational connections

There were 15 performances of *Nanay* between November 25 and 30, 2013, and 431 people came to see them with PETA drawing on their contacts in universities, schools and the theatre community.[23] We were able to get members of a range of non-governmental and governmental organisations to the play, no small undertaking given the challenges of travelling through Manila traffic.[24] Eight members of six families of domestic workers in Vancouver attended. Most flew from other regions or islands: Baguio, Cebu and even Leyte (just days after the latter was hard hit by typhoon Haiyan/Yolanda), taking time from work and making a great effort to come to Manila. Just as remarkable was the global extent of expertise in the audience. Each performance culminated in a facilitated public forum, and in these activists spoke from their experiences as organisers in South Korea and Saudi Arabia, as well as at home in the Philippines. Researchers in the audience brought into the conversation their research in Italy, Malaysia, New Zealand, Singapore and Spain, as well as among children of OFWs in the Philippines. Audience members testified to their time as an OFW, as a fertility nurse in Saudi Arabia, as migrants to Canada and as an undocumented worker in New York City. They spoke of mothers and grandmothers working in Canada and the United States, and of friends working as OFWs in Hong Kong, Rome, New Zealand, Dubai and Holland. At least one person spoke from their own experience at each talkback, and in one forum, personal stories were told from eight different countries. As Aguilar notes, the Philippines itself is a transnational terrain: "the figure of the migrant worker breaks down the borders of inside and outside, even as it links the origin and destination."[25]

Many in the audiences confirmed that they knew little about Canada and were interested to learn more. They spoke to the idealisation of Canada as a migration destination and *Nanay*'s disruption of it. From three different audience members we heard:

> I was trafficked in United States of America. Filipinos have that connotation that when you go to Canada or the United States of America it's like the dream come true. When you put your feet on the land it's like heaven. But they don't actually think about the actual or the other side – the negative side of what could happen to you as an OFW worker.
>
> So, actually I was very surprised to find out that there's so many issues, when I thought all this time that – comparing Hong Kong and Canada – Canada had better working conditions. I have a cousin who is still in Canada. She worked in Hong Kong before moving to Canada and she was saying that the conditions in Canada were so much better compared to what she suffered in Hong Kong. Again, [the play] was a surprise. But at least, thank you for bringing this up, that there are still issues with migrant workers no matter how rosy the picture may look.
>
> We're always confronted [...] [Switches to Tagalog] maybe many of us know or have family or relatives who are abroad. And we're always thinking that if you're outside of the Philippines, life is good. From the play, it makes clear that life outside of the country isn't all rosy. Whatever people are experiencing in the Philippines, it is heavier for those who go abroad because they don't have people they can depend on there. [Switches back to English.] So, things are not what they seem. I think we are cornered with the conclusion that when you're outside

> the country, you'll just feel very liberal in a way. There's snow. And it's hot here. So, things are not what they seem and I hope that the other audience members realize [switches to Tagalog] that they shouldn't get cornered into that way of thinking.

Family members of PWC of BC members in Vancouver who attended the play (and were interviewed afterwards) confirmed that even they had limited information about the lives of their mothers and sisters in Vancouver. Carlos, whose mother left when he was seven years old, and had returned only twice (once when he was 16 and again when he was 20 years old), was asked whether his mother ever shared her experiences as a domestic worker. "Sometimes she shared," he said, but "I know she's hiding something. She doesn't want us to feel bad also. But of course, I know. I know the nature of her work."

Other family members spoke of a slow unveiling of details as they grew older. In Grace's case, this happened when she became an OFW herself, working as a nurse in Saudi Arabia. Her sister left when she was still in elementary school. She doesn't recall her sister telling her anything negative about working as a caregiver in Canada: "I didn't know anything about the reality of what's life like being a migrant in [Canada]. So I was even thinking that everything is okay, everything is nice." And "when they come home, you think they're rich. They go here and there but you don't know that they just borrowed that money. When you see their hands, you think they have allergies but really it's just because they always clean the toilets." Even when she was in college, Grace recalls being "one of those persons, frankly speaking, taking for granted our family working abroad. Like I mean: 'I need money on this date.' Wow, when I go working [as an OFW in Saudi Arabia], I said [to my family asking for money] 'Oh my god, you need to wait for many days for me to make this money.'"

Grace reasons that her sister did not tell them the truth of her life in Canada because "she is worried that maybe we will worry about her."

> Like [...] if she will tell the real situation there, *baka yung*, maybe my father and mother will get worried about her and maybe she's thinking about the education or the financial situation for the family. So she's in the hot spot. She cannot say, 'I'll go home.' When you go home, what will you do? Your family will not have something to eat. Really at that time when she went abroad, we were very... life was so hard at that time really. [...] You do not leave for yourself, you know. You love your family, you help. But I think it's too much. At least I did my part. I finished school. My sister, she will do everything for the family. Look, I'm crying!

Audience members spoke of the ways that *Nanay* expanded their understanding of the lives of domestic workers in Canada. Though Carlos already had intuited that his mother shielded him from negative aspects of her job, and Grace and other family members interviewed after the play said that they had learned more of their sisters' experiences as they had grown older, they spoke of the ways in which the play solidified this knowledge. From Carlos, "After watching the play it was like I was able to picture the nature of my mom's work, my mom's relationship with her employer, see how she adapted. I was really a bit saddened by that. But I know it's the real deal, you know? That's the real thing. [Long pause.]" From Glecy,

> I liked Joanna [a monologue of the domestic worker in *Nanay* who tells of her difficult employment situations]. Not really liked but was touched. *Nakarelate ako. Nakarelate* because *di ba* this Joanna, she worked, [like my sister] for employers who did not like her or treat her well [...]. [My sister], she also experienced that and I am seeing it in my own eyes [in Joanna's scene]. And Ligaya [the monologue of the domestic worker who tells of her reasons for leaving Manila], she left for her son, right? For her children. [My sister also] left because she has to. I don't know if she has to but she left because she wants to help us. To send us to school. [...] [We are now] a lawyer, pharmacist, nurse and psychologist: we are all professionals because of her help. What she sends us is our tuition fees, which is much money. So yah. But now life is easier for her. Ay, no! That's not the case! She's also in great debt in Vancouver because she helped my sister have her own pharmacy. So thanks to Canada. [Laughs.]

The play was not only a vehicle for transporting stories across oceans and generating new witnessing relationships within transnational families, but a means of sparking trans-oceanic conversations within and beyond transnational families. When Glecy spoke with her sister after seeing *Nanay*,

> [My sister] was actually laughing, not laughing at me, but the way I talked to her about it [the play]. When I told her: 'Well you know it's all true,' she laughed. I don't know why she laughed. But she said 'Yah, I told you everything about those things.' Because it's actually [...] she tells us about those things. It's not really that new to us [these] issues about working abroad. It's not really that new. So yah, I told her I know it's true and she said 'Yah, I know, I told about those, right?' That's what she said. And then she was laughing. It's not laughing about those experiences. I think she was laughing because I was the one telling her those things [about life under the LCP].

When we met up with Glecy's sister in Vancouver several months later, she was still amused that Glecy had assumed the role of expert on the LCP after seeing *Nanay*. But she also noted that Glecy is now more interested in and concerned about her life in Canada. When they speak to each other, Glecy asks more questions – and more specific questions – about her work and life in Canada.

The talkbacks were themselves occasions that brought the issues raised in the play very close to audience members and allowed organisations and academics to connect with each other and with audience members who they might not otherwise reach.

> Hi. I'd just like to say for me the play hit really, really close to home. Especially the child of the domestic worker. My mom has been working in Canada for two years now, except she's not a domestic worker. She's been working in some restaurant as a cook for two years now. What the child described is pretty much exactly what I'm going through right now, except I'm in a more middle-class situation. She wants me to *ano*, she wants to bring me and my sister over. She's apparently [...] She's trying to file her permanent resident application, which is going to be approved soon apparently. She wants to bring me over there in the middle of the school year, like this January. I'm going back to high school there. I'm a first-year college student. I've got a scholarship in my university. I have really high hopes. I really dream to finish my course here and probably get a Masters and move to Canada later or something. But not now. But even then, I feel – I know that my mom busts her ass off working ridiculous hours – 9 to 8 or some shit, 6 days a week. And I don't know, I just don't really have the heart to tell her that I don't want to go to Canada. And my sister feels the same way. But I don't know, whenever we talk to each other, she always says how good life is there in Canada, how lovely the people are. I guess she doesn't tell me the bits about her working ridiculous hours. I'm not sure how much she tells me is sugar-coated. I just really don't know how to feel about this. I'd just like to get that out.

As friends from university comforted him by putting their arms around him, May Farrales, a Filipino-Canadian with a long history as a youth organiser and former director of the PWC of BC who was co-facilitating the talkback, spoke, drawing on her MA research with Filipino youths in Vancouver high schools.[26]

> I appreciate you sharing that. I think that's a conversation that doesn't happen often. For the students that I was working with, a lot of them expressed a lot of anger over having to go back to high school. Others talked about being surprised by having to prepare to leave [the Philippines] right away. From others who have worked through the process I hear them encouraging people in your situation, as best as they can, to open communication with their mothers or their parents about how they're feeling. And, also, what I heard from them is that it helps to seek out support from other students that have gone through

> similar experiences, to be able to compare experiences and to compare how people have coped through these very drastic changes in their lives. I don't have any particular advice but that's what I heard from the students who are now in Canada and who went through the process that you're about to embark on. When I asked them: 'What kind of advice would you give to a friend of yours that was going to migrate to Canada in one month's time? What would you tell them?,' they said stuff like: be prepared to go through re-meeting your mom and living with her. One young man suggested talking to her and telling her about how he felt. Another youth shared that they would have liked to have known that they were going to get pushed back into high school. They also would have liked to have known that they would be going through the English as a Second Language program. They also wanted to know about how to get in touch with other Filipino youth building support for each other.

To which the young man in the audience responded, "I'm not going out without a fight. I'll try to... I hope I'll be able to reach a compromise. I just hope I'm still here next year. I have so much here. I don't want to let all that go. Thank you." Fifteen minutes later, another woman spoke partially in response to the previous exchange, identifying herself with an organisation that works at the CFO in Manila, where the pre-departure orientation seminars are held. She offered:

> I know the US case. I'm not so familiar with how it works legally in Canada. In the US case you could postpone your visa application until after you graduate. So, you could keep telling the US government that you're a student and you'll come in when you're finished.

Changing the conversation

But beyond an exchange about the struggles of OFWs and their families in or going to Canada, the talkbacks became a corrective and supplement to the play that staged different political questions and re-centred the conversation on the Filipino diaspora and the Philippine labour brokerage state.[27] "I found that last scene a little bit frustrating," we heard. The audience member expanded:

> [The play] starts out with these very personal narratives. Then it ends with these two representatives of the government. The play felt like it was still-lives of different people and I didn't get any really sense of the possibility of change. I felt like people were stuck. I'm kind of excited about the possibility you were talking about of how people are using this program as a tool for family reunification. And what's the relationship of the people having positive experiences [in the LCP] to the people

> who are in the more oppressed situations. I'm interested in what these things do for Filipino migrants' community formation. How are people using community formations to support themselves? Everyone just felt very detached. [Let me be clear.] When I'm talking about these positive experiences and family reunification, I'm interested in how any community formation that comes from them might be in direct opposition to people who are in more oppressive situations. So, you might have community formations within the Filipino diaspora that are in antagonistic relationships to one another. I was frustrated with the ending because it went into the government representatives. I already know that I'm supposed to question these people. Like that seems to be playing into [...] like the Canadian government says 'you know, you stay here for 2 years, you can become one of us,' and it's trying to incorporate some of these people into the Canadian community in ways that severs them from the direct ways that people in the Philippine diaspora are being oppressed. So, it's creating this divide, or it's kind of breaking the Filipino diaspora up, or it's kind of frustrating any attempt at real community formation.

These questions, implicitly about class fragmentation and co-optation within the Filipino diaspora (foreshadowing the discussion in Chapter 4), played out in complex ways within the talkbacks in which predominantly middle-class audiences troubled about their relationships to both domestic workers and overseas migration. One criticism of this production – that PETA Theater Center is essentially a middle-class venue[28] inaccessible to many labour migrants – was both on and off target (another response to it will be explored more fully in the next chapter). That the audiences were largely middle class there is no doubt. The middle-class status of many in the audience is suggested by the fact that so many of the stories of immediate family members were of mothers and grandmothers in the United States and Canada. However, given that Canada sits at the top of the migration hierarchy, one might argue that PETA is a very appropriate venue for *Nanay*, because this is the audience most likely to migrate to Canada or to have relatives who have migrated there.

In some ways similar to the Vancouver production, the play also reached the audience as employers of domestic workers. From one audience member,

> For me, what came to my mind was domestic workers not in the international setting but in the local setting. [...] We're [also] the ones who are the employers and the nannies would be the people living with us. I'm thinking that this situation starts with us. How many of us have domestic workers at home? I think a lot of us would have domestic workers at home and this issue of having domestic workers should be tackled more on the local setting. I can relate with the employers a lot – like sometimes I feel that the *kasambahay* are not doing what I want and I feel angry. What [the employers in the play are] feeling is sometimes what I'm feeling.

A second audience member responded:

> Can I just add? I totally appreciate that point. The Philippines is the 2nd signatory to the rights of domestic workers internationally. It's because they're trying to protect their domestic workers abroad. There's a law in the books for the Philippine context for the rights of domestic workers [...] But nobody really enforces that. The Philippine government is aware of the issue and they've passed the laws. But there is no movement to implement the law in the local context.

In another forum, an audience member said:

> It feels like I have a very cold instrument stuck in my body right now. Because it's like watching the stories of domestic help which we come home to every night in our own homes. I was really wondering what the domestic helpers, our own domestic helpers, would see or feel if they watched something like this. Because as far as I know if they come from the rural areas, it's not really work, but it's work with an opportunity. We call them *katulong*. It's like an extension of our homes, but not really an extension of the home. So, it's really a very different discourse all together. And it's very interesting how artists from other countries have given us a new take into something which is right on top of our noses. Something which we don't really see as very problematic. Because we look at the domestic help situation differently. It's not really like a job. It's a very complex discourse of having to find a job and looking for one in the city and having the opportunity to be part of that city. And it's even better if you go abroad. I cried with Ligaya [character in play] because I can relate to her as a mother, but it's really like there is a tug of war between looking at her as opportunity and looking at her as a tragedy.

Beyond direct experience as or with a migrant domestic worker, the figure of the domestic worker permeates national discourse. "Domestic workers' bodies," Neferti Tadiar has written, "serve as objects of the nation's struggle for subject-status on the global scene."[29] Audience members raised concerns about the ways that domestic workers are devalued by middle- and upper-class Filipinos, along with the shame of being mistaken for a domestic worker. From three different audience members:

> I guess what's sad about some of the stories is that the dignity is somehow diminished or demoralized. I know a story of a child of a Filipina who was working in Hong Kong for Oxfam. It's an international NGO but her child [who was living with her mother there] was ashamed to say she was a Filipina because the image is that if you're a child of a Filipina, you're a child of a domestic worker. And she was ashamed of that.

I think that maybe subconsciously Filipinos are racist to themselves. Maybe there's a subconscious thinking that we're not really an important people in the world. [...] These little things like taking care of the elderly, taking care of the children: it's something that's actually a big contribution that people should really uplift and see it for the great work that it is.

I find similarities [to the attitudes of Canadian employers] among the attitudes of the landed Filipinos and the othering of their fellow Filipinos. At the airport you would see them regarding the domestic workers and the caregivers with a different eye. These caregivers, they have Louis Vuitton bags, but the snooty behaviour [of other Filipinos] is still there. Actually I see this play as more targeting the snotty Filipinos [laughter], not the relatives of the workers because they know – *nakakarelate sila tsaka alam na nila yung* [they can already relate since they already know]. But I think *yung hindi na address yung* attitude of indifferent Filipinos towards this set of Filipinos abroad.

Numerous Filipino scholars have written about the shame triggered by Filipina domestic workers among Filipino elites and upper and middle classes, on both a personal and national level.[30] Like the child of the Oxfam employee mentioned above, there is discomfort about being read internationally as an OFW and a concern that the nation's reputation is tarnished by deploying so many Filipina women as domestic workers. Although Aguilar notes that the emotion of shame is not as prevalent as it was in the mid- to late 1990s, during and immediately after the crisis of Flor Contemplacion's execution, he argues that it "continues to be a strong strain of thinking" and it was apparent in the talkbacks after *Nanay*.[31]

Perhaps most pressing were the concerns expressed by middle-class audience members about being unwillingly absorbed into the machinery and necessity of migration. This was felt intensely by a class of medical students attending the play:

AUDIENCE (Female student 1): Hi, I'm a medical student but my premed is nursing so I'm a nurse. I got pissed when one of the characters said "Nanny, nurse – *pareho lang yun.*" I mean seriously: they are not the same. I graduated 2012 from nursing. I took up Med right after. Some of my batch mates from nursing went to the States, some of them went to UK, some of them [...] anywhere [laughs] basically just to land a job. But most of my batch mates went to Med because most of the nursing graduates cannot find a job. So we resorted to Med. That's what's happening now. They said that the trend is going to go backwards one of these years because so many of those nursing students who can't get a job are going into Med. And then there will be a lot of Med students and they are going to have to land a job elsewhere. So I just got pissed with that comment because I studied for 4 years, I worked my ass for 4 years just to get a license and pass the board and try to top the boards. And then what? You're just going to call me a nanny? Hell no! [laughs]. I got pissed. And then looking at my classmates now who are working abroad, I don't see them as nannies. I mean they're professionals. And so it's kind of saddening for me to see that some nurses are resorting to nanny work because that's not what we're trained to do. We're trained to help patients, not people who can't take care of their own children.

AUDIENCE (Male student): About the question about the government encouraging the migration of workers. What I can say about that is the government does not explicitly encourage migration but a lot of the structures the government has right now coerce them or force them to work abroad. There's no particular comment from the Department of Labor and Employment on this matter of nurses being domestic workers. Like they're always branded as the 'New Heroes' or the 'Bagong Bayani.' And the thing about that: it kind of clouds our judgment because we mask them as, at least the media masks them as heroes. So I don't know. For the government to say that they are not encouraging migration: they have a pretty good way of forcing people to work for other countries.

AUDIENCE (Female student 2): So, to add, it kind of speaks to the way we've adopted this neoliberal global model in which migration alone is a way of validating everything you've studied and everything you've learned. It's like 'Okay, I've done everything that I've could, and then they're accepting me in a first world country – yay, I guess I've made it.' I think the glamourizing narrative of 'bagong bayani' has shifted to that instead. It's not just about becoming a hero but becoming a hero who is internationally recognized to some strange degree. And to me it's just so weird that we have all these women who are experiencing a very cosmopolitan way of life, and yet they are confined to the domestic sphere.

What emerged was an insistence on the right to stay in the Philippines. One strong political impulse that emerged was not to repair the LCP but rather to stop the flow of migration: "It really pains me. In fact, my parents are in the States. Many years they said: 'we'll petition you.' I said no. I said no. [...] For me, my fear is if the LCP is made very good, I'm afraid [that more of my] countrymen [will] be leaving the country. It really pains me."

Perhaps even more disruptive to Canadian perspectives and preconceptions, was the rejection of Canada as one of the most coveted of migration destinations. The LCP is commonly legitimated within Canada through comparison. Few in Canada would dispute the hardship of family separation and a minimum of two years of indentured servitude required by the LCP. But the pay-offs that come with permanent resident status and the opportunity to sponsor one's dependents to Canada are thought to counterbalance and even outweigh these hardships. Saudi Arabia and countries in the Middle East are considered to be the most dangerous and most exploitative destinations – a world away from Canada. And so, it was truly disruptive for Grace, a family member coming to see *Nanay* in Manila, to tell her story of twice refusing the LCP. In 2009, her aunt in Canada found her an employer. Grace, a recent nursing graduate, had signed the contract with the employer in Canada and was three months (halfway) through the required caregiver programme (taken in the Philippines). She backed out of the contract because she decided that she could not leave her daughter, who was young and suffering from asthma. In 2011, Grace went to Saudi Arabia instead to work as a nurse and stayed for 25 months. Her experience there was "great." Going to Saudi Arabia, she said, was "a gamble" and her sisters from Canada and Australia had called her parents when they first heard of her plans: "They were really mad at me." Their concern was that the hospital she was hired to work at was "maybe a ghost" and she would find herself working under poor conditions as a housemaid. It was while Grace was in Saudi Arabia that her sister in Canada first revealed that she herself was essentially working as a housemaid in Canada. Since returning from Saudi Arabia, Grace had once again been considering migrating to Canada. Asked whether the play had changed her perspective, she responded, "Even my decision! I was even planning to go to Canada through the LCP. I was telling my sister [after seeing the play]: 'Wow, I love you because you sent me to this play and now I am very enlightened.'" Instead of migrating to Canada under the LCP, Grace had decided to continue to practice her profession as an OFW nurse in Saudi Arabia. In the cosmic gamble of self-making and fate playing,[32] after seeing the play, Grace reassessed the stakes of permanent migration and the loss of professional standing in Canada against the risks of temporary migration as a nurse in Saudi Arabia. That the dice rolled in favour of Saudi Arabia is almost unimaginable to most Canadians, who tend to take as self-evident that "the world needs more Canada."[33]

Talking back to *Nanay*

The insufficiency of our efforts to "transnationalise" the play was laid bare when an audience member said plainly, "It was obvious to me [that the actors] were talking to white people." Because they were speaking in English, "they were talking like we didn't know what they were talking about."

Talking to Filipino audiences like they "didn't know what they were talking about" calls up different histories, violences and sensitivities, and English is at the centre of many of them. English, Renato Constantino wrote, "was the beginning of [Filipinos'] miseducation, for they learned no longer as Filipinos but as colonials."[34] As an indication of the scope of this "miseducation," at the end of the Spanish colonial period (which spanned 1521–1898), only 2.46% of an adult population of 4.6 million Filipinos (mostly elites) spoke Spanish, but by the end of the American colonial period in 1935 (after only 37 years), 26% of Filipinos of a total population of 16 million spoke English.[35] The "English effect," Constantino argued, was two-fold: it led to an idealisation (and amnesia around the violence) of U.S. colonialism and situated English as the marker of and gatekeeper to resources and privileged status. More specific to the Philippines labour export policy and the LCP in particular, proficiency in English has been central the Philippine state's marketing of the national population as high-value labour exports. In 1974, the Philippine government advertised Filipino labour globally on the basis of English proficiency, declaring in a *New York Times* advertisement that "Our labor force speaks your language."[36] This state rhetoric continues today,[37] with the effect of creating Filipino domestic workers as "a kind of labor aristocracy in the regional and global domestic service industry."[38] Certainly, English proficiency is foundational to admission to Canada, where proficiency in one of the two national languages is required for a temporary work visa through the LCP.

Filipino domestic workers bringing English back from Canada to the Philippines (as we effectively scripted the characters in *Nanay* to do) carries additional meanings. Beatriz Lorente writes about the careful monitoring by middle-class and lower middle-class OFWs of their use of English when they return home, in an effort to avoid being criticised for putting on airs.[39] The English language monologues by domestic workers who have survived the LCP, gained permanent resident status and sponsored their families to Canada are even more fraught. This is because they have crossed the line that divides OFWs on temporary labour contracts from *balikbayans* (immigrants to North America who return to the Philippines intermittently). In contrast to OFWs, who are cast as "national heroes" (for the remittances they send home for their families' survival) and as objects of pity (because of the deprivations and humiliations they endure as migrant workers), it has been argued that *balikbayans* tend to be viewed as collaborators with U.S. colonialism, who are "steeped in their own sense of superiority" and whose permanent departure amounts to a kind of betrayal of the nation.[40]

The inherent impossibility of translation from one language to another is a different but equally important concern. A repeated criticism was that English cannot communicate the same meanings with the same intimacy as would be possible in Tagalog.[41] We had our first inkling of this when we worked with a Filipino-Canadian actor, Lissa Neptuno, in Vancouver, to develop a new domestic worker monologue. When we first drafted (but did not use) the monologue for the Vancouver production we had translated it fully to English. For the Manila production we returned more and more of it to Tagalog and Taglish, to the mixture of the languages used in the original interview. The actor workshopping the monologue noted that the mode of storytelling changed as the narrator moves in and out of Tagalog, English and Taglish. Not just the words but the mode of storytelling. The storyteller was more likely to use English, for instance, when she answered a question as "a research subject." The Filipino actors in Manila articulated the same point about the impossibility of translation and the intimacy of Tagalog. Marichu Belarmino felt that her character, Ligaya, would have had a different emotional resonance if her words were spoken in Tagalog: "When she's saying 'lives here are difficult, but I promise myself I would just go on. I just have to fight,' when you say that in Filipino, I believe it has a stronger message." Another Filipino actor, Joanna Lerio, spoke about how "weird" she first found it for the Filipinos in the script to be speaking English. She accepted that much of the script was in English because it was verbatim and the Filipinos in Canada had been addressing Canadian interviewees. But she wrestled with the fact that the audience would be Filipino. "So, it was hard for me to find the balance, and I was just happy that our director permitted me to play with the language and switch between Filipino and English." The use of English was nonetheless hard to resolve "because I imagined the audience, how they would take the stories and relate. [From the talkbacks we know that] some of the audience felt detachment from the characters. Like, I know that she's a *nanay* character but I can't feel that she's my *nanay*. Something like that." As an actor, the use of English made her feel that that she holding a secret: "this part is too formal or I was hiding something. Like, I will not tell you everything." Coming to Manila to tell some of what domestic workers are reluctant to share with their families, the use of English inadvertently communicated that something was being left unsaid or withheld.

A Western framing may unwittingly be even more deeply embedded within the play. A theatre professional attending the play also noted that the use of English made it "glaring for me is that the text is written for a foreign audience." For Filipino audiences, he said, "it's kind of alienating. You know that it's supposed to appeal because all of those stories are familiar. Here in the Philippines, we are all aware of it. I mean, I grew up in a family wherein both my parents left. But somehow it wasn't really resonating with me." What resonated more for him were the employers, not because he had personal experience as an employer but because "there's something about the text." He wondered whether the employer texts were verbatim because he detected a difference from the domestic worker monologues. "There are words jumping out of the monologues of the employers that made me wonder why they resonated." It had something to do about the way "it was arranged: the words, the way it was performed. The employers appeal more to the intellect. And the workers were more emotional. And in that way, there's sort of an imbalance. I don't know if that's the goal. [...] The monologues [of the domestic workers] are somewhat emotional but there's something missing. Between [the domestic worker monologues,] what the employers want to say [and] the play itself, I feel that there is some sort of an imbalance. Maybe because one was intellectual, the other one was emotional: I don't know if they meet." In fact, the interviews with employers and domestic workers were done under very different conditions: the former as formal interviews with a university researcher, the latter as testimonials with activists from the PWC of BC (or in the case of the youth, with fellow Filipino youth). In a deep methodological sense, they do not meet. A theatre professional had the ear to hear this, and he both registered and upended Western binaries when he found resonance in the more intellectual, less emotional text.

Lingering in a transnational conversational space

Our privilege and distance from the issues in the Philippines was evident right from the onset. We arrived in Manila to begin rehearsals for *Nanay* on the morning of November 8, 2013, just as Typhoon Haiyan (Yolanda) was expected to hit the metropolis. Schools were closed and the roads, usually clogged with traffic at 9 am, were eerily empty. The winds picked up velocity throughout the day. We had entered into the edges of the strongest tropical cyclone on record, the most powerful typhoon ever to hit land and the deadliest in Philippine history: one approximation is that 6,300 people died. A state of emergency was declared and UN officials estimate that over 11 million people were affected, with many left homeless. Colleagues at the UP Diliman campus in Manila were numb; in the early days following the typhoon, many of the faculty members at their sister campus in UP Visayas, Tacloban (one of the areas most devastated by the typhoon), were presumed to be dead. Filipino friends and family in Canada, so many of whom come from the Visayas region, were desperate to receive information about their loved ones. We and the play went on; our privilege is a given.

And yet, remarkably, the play did open a transnational conversational space in both intended and unanticipated ways. It transported stories and lived experiences from Canada that disrupted its relative innocence at the top of the migration hierarchy, as well as stories that opened a more intimate set of exchanges and understandings among members of transnational families. It led to different (and new to us) conversations: the particularity of discourses about domestic workers in Manila, possibly tied not to the family but the house and city; and to consideration of how migration policy and programmes in specific countries affect community formation within the Filipino diaspora. The play – from delivering invitations to talkback fora – brought us face-to-face with other sites where the violence of migration is enacted: for instance, at the Canadian embassy; among university students and highly educated professionals who anticipate and resist being absorbed within the forces of migration. The uneasy travels of the play left us, moreover, more fully attuned to the inevitable incompleteness of our translations and the need to do more work to better understand our place within the contemporaneity of colonialism.

Notes

1 See Anju Mary Paul, "Stepwise International Migration: A Multi-Stage Migration Pattern for the Aspiring Migrant," *American Journal of Sociology* 116, no. 6 (2011): 1842–1886; "Capital and Mobility in the Stepwise International Migrations of Filipino Migrant Domestic Workers," *Migration Studies* 3, no. 3 (2015): 438–459; *Multinational Maids: Stepwise Migration in a Global Labor Market*

(Cambridge, UK: Cambridge University Press, 2017); Aida Geraldina Polanco Sorto, "Behind the Counter: Migration, Labour Policy and Temporary Work in a Global Fast Food Chain" (Unpublished Dissertation, University of British Columbia, 2013).

2 Oscar V. Campomanes, "New Formations in Asian American Studies and the Question of U.S. Imperialism," *positions* 5, no. 2 (1997): 523–550; see also Robyn M. Rodriguez, "Toward a Critical Filipino Approach to Philippine Migration," in *Filipino Studies: Palimpsests of Nation and Diaspora*, eds. Martin F. Manalanson IV and Augusto F. Espiritu (New York: New York University Press, 2016), 33–55.

3 Andreas Wimmer and Nina Glick Schiller, "Methodological Nationalism and Beyond: Nation-State Building, Migration and the Social Sciences," *Global Networks* 2, no. 4 (2002): 301–334.

4 Gayatri Chakravorty Spivak, "Thinking Cultural Questions in 'Pure' Literary Terms," in *Without Guarantees: In Honour of Stuart Hall*, eds. Paul Gilroy, Lawrence Grossberg, and Angela McRobbie (London and New York: Verso, 2000), 335.

5 Filomeno Aguilar, *Migration Revolution: Philippine Nationhood and Class Relations in a Globalized Age* (Singapore and Kyoto: National University of Singapore Press in association with Kyoto University Press, 2014), 17.

6 Beatriz P. Lorente, *Scripts of Servitude: Language, Labor Migration and Transnational Domestic Work* (Bristol, UK: Multilingual Matters, 2017), 27.

7 Ibid., 54. See also Anna R. Guevarra, "Supermaids: The Racial Branding of Global Filipino Care Labour," in *Migration and Care Labour: Theory, Policy and Politics*, eds. Bridget Anderson and Isabel Shutes (Houndmills: Palgrave Macmillan, 2014), 130–150.

8 We thank Michiyo Yonenon-Reyes of the University of the Philippines Diliman for these observations.

9 Caroline Sy Hau, "Privileging Roots and Routes: Filipino Intellectuals and the Contest over Epistemic Power and Authority," *Philippine Studies: Historical and Ethnographic Viewpoints* 62, no. 1 (2014): 29–65; Martin F. Manalansan IV and Augusto F. Espiritu, "The Field: Dialogues, Visions, Tensions, and Aspirations," in *Filipino Studies: Palimpsests of Nation and Diaspora*, eds. Martin F. Manalanson IV and Augusto F. Espiritu (New York University Press, 2016), 1–11. Given commitments to "public outreach" and "impact" in Anglo-American academic contexts, we were each able to access funding from the major national agencies in our respective countries, expressly to take *Nanay* to the Philippines. Our activities in the Philippines were supported by funding from the United Kingdom's Economic and Social Science Research Council, and the Canadian Social Sciences and Humanities Research Council.

10 Rick Bonus, "'Come Back Home Soon': The Pleasures and Agonies of 'Homeland' Visits," in *Filipino Studies: Palimpsests of Nation and Diaspora*, eds. Martin F. Manalanson IV and Augusto F. Espiritu (New York: New York University Press, 2016), 390.

11 See Phanuel Antwi, Sarah Brophy, Helene Strauss, and Y-Dang Troeung, "'Not without Ambivalence': An interview with Sara Ahmed on Postcolonial Intimacies," *Interventions: International Journal of Postcolonial Studies* 15, no. 1 (2013), 113–114.

12 Eugene van Erven, *Stages of People Power: The Philippines Educational Theater Association* (Verhandelingen no. 43, The Hague: Centre for the Study of Education in Developing Countries (CESO), 1989), 74.

13 Eugene van Erven, *Community Theatre: Global Perspectives* (Routledge: London and New York, 2001).

14 For an account from the director, see https://pushfestival.ca/manila-blog/.
15 Non-verbatim text is italicised.
16 Because the actor hired to perform this scene in Manila found she was unable to fit the rehearsal period into her already busy schedule, this scene was dropped from the PETA production. For the script of the full scene, see Caleb Johnston and Geraldine Pratt, "Taking *Nanay* to the Philippines: Transnational Circuits of Affect," in *Theatres of Affect*, ed. Erin Hurley (Toronto: University of Toronto Press, 2014), 192–212.
17 Geraldine Pratt with Migrante British Columbia, "Organising Domestic Workers in Vancouver Canada: Gendered Geographies and Community Mobilization," *Political Power and Social Theory* 35 (2018), 99–119.
18 We thank Jessica Hallenbeck for making this video recording.
19 For more on Migrante BC's advocacy work, see www.migrantebc.com.
20 This erupted as an issue in the Canadian media in 2014, when the LCP was criticised for having morphed into a hidden family reunification programme. The Employment Minister of the day, Jason Kenney, claimed that the LCP had "mutated" into a family reunification programme "whereby thousands of foreigners are coming to work for their own relatives in jobs that might not otherwise exist" (Jennifer Hough, "Canada's Live-In Caregiver Program 'Ran Out of Control' and Will Be Reformed: Jason Kenney," *National Post*, 24 June 2014, online; see also Bill Curry, "Live-In Caregivers May Be Next Target of Immigration Reform," *Globe and Mail*, 23 June 2014, online; Citizenship and Immigration Canada, "Immigrant Overview – Permanent Residents," Canada Facts and Figures, Ottawa, 2014; Douglas Todd, "The Caregiver Conundrum: Caregiver Plan Popular, Problematic: Nine Simmering Debates about Canada's Approach to Foreign Domestic Workers," *Vancouver Sun*, 24 May 2014, A1, A6.) Note was made of "a private report from the Canadian consulate in the Philippines … [that] pegs the portion of Filipino caregivers working in Canada for their own relatives at 40 to 70 percent" (Douglas Todd, "The Caregiver Conundrum: Caregiver Plan Popular, Problematic: Nine Simmering Debates about Canada's Approach to Foreign Domestic Workers," *Vancouver Sun*, 24 May 2014, A6). Employment Minister Jason Kenney offered as proof his presence at a seminar "on nannies' rights" in Manila "a few years ago" where all 70 LCP registrants in the room were to be employed by their relatives in Canada, and "none had any questions about rights." "All they wanted to know was: what was the penalty for working outside the home illegally, and how long would it take to sponsor family members" (Jennifer Hough, "Canada's Live-In Caregiver Program 'Ran Out of Control' and Will Be Reformed: Jason Kenney," *National Post*, 24 June 2014, online). In an article published on the front page of the *Vancouver Sun*, an immigration lawyer is quoted as stating that when live-in caregivers are employed by their own families, it is difficult to know whether they are "pulling a fast one" or whether the foreign domestic worker is "properly trained" or in fact "performing their duties" (Douglas Todd, "The Caregiver Conundrum: Caregiver Plan Popular, Problematic: Nine Simmering Debates about Canada's Approach to Foreign Domestic Workers," *Vancouver Sun*, 24 May 2014, A6). A University of British Columbia Professor Emeritus, Prod Lacquian, "who was raised in the Philippines and has written about Asian immigration" (but notably not the LCP), was quoted as recognising that "'the vast majority try to get out of the LCP as soon as they can,' occasionally to work as live-out helpers, but mostly to pursue careers in the health care industry" (Douglas Todd, "The Caregiver Conundrum: Caregiver Plan Popular, Problematic: Nine Simmering Debates about Canada's Approach to Foreign Domestic Workers," *Vancouver Sun*, 24 May 2014, A6). LCP registrants are not bound to domestic care-work

after completing the programme and so this comment is irrelevant, but it is deployed to suggest an abuse of the programme. The main concern at the time was that labour market demand in Canada for live-in caregivers was illusory and that Filipino families were effectively "gaming" the immigration system by using the LCP as a means of gaining entry for their low-skilled relatives who would not otherwise qualify as legitimate immigrants. The gendering and undervaluing of care work as unskilled was unquestioned and doubts about the fraudulent use of the programme were insinuated rather than substantiated. That Filipino-Canadians may be in need of child or eldercare – from family members or otherwise, was not considered or investigated.

21 Nina Glick Schiller and Ayse Çaglar, "Migrant Incorporation and City Scale: Towards a Theory of Locality in Migration Studies," *Journal of Ethnic and Migration Studies* 35, no. 2 (2009): 177–202.

22 In retrospect, it seems odd that we harboured illusions that these invitations to top bureaucrats and diplomats would bear fruit. Would we have expected this kind of access to top Canadian bureaucrats? We might have realised that the interest of the tour, as Migrante organisers undoubtedly understood, was in experiencing the faces (and facelessness) and the disregard of the migration bureaucracy.

23 Faculty members and students came from Ateneo de Manila University, University of Sto. Tomas, University of the Philippines Diliman, Polytechnic University of the Philippines, University of the East medical school, Manuel Luis Quezon University, and Miriam College. Theatre audiences included PETA staff and artist-teachers, members of Teatro Balagtas, MAPUA Institute of Technology Theater Group, the Far Eastern University Director of Culture and Arts, PETA-Metropolitan Teen Theater League youth, and the ENTA Theater group.

24 ECPAT Philippines (focuses on Trafficking of Children), Batis Center for Women, Philippine Migrant Rights Watch, Scalabrini Migration Centre, Migrant Forum in Asia, National Council of Churches, Catholic Bishops Conference for Migrants, Migrante International, KPD (Democratic Peoples' Movement) and the Philippine Commission on Women.

25 Filomeno Aguilar, *Migration Revolution: Philippine Nationhood and Class Relations in a Globalized Age* (Singapore and Kyoto: National University of Singapore Press in association with Kyoto University Press, 2014), 2–3.

26 May Farrales, "Delayed, Deferred and Dropped Out: Geographies of Filipino-Canadian High School Students," *Children's Geographies* 15, no. 2 (2017): 207–223.

27 Robyn M. Rodriguez, *Migrants for Export: How the Philippine State Brokers Labor to the World* (Minneapolis: University of Minnesota Press, 2010).

28 The criticism of "middle class-ness" likely extended beyond the specifics of the appropriate audience for our production to PETA itself (Eugene van Erven, *Stages of People Power: The Philippines Educational Theater Association* (Verhandelingen no. 43, The Hague: Centre for the Study of Education in Developing Countries (CESO), 1989)).

29 Neferti X.M. Tadiar, "Domestic Bodies of the Philippines," *Sojourn: Journal of Social Issues in Southeast Asia* 12, no. 2 (1997): 165.

30 See Filomeno Aguilar, *Migration Revolution: Philippine Nationhood and Class Relations in a Globalized Age* (Singapore and Kyoto: National University of Singapore Press in association with Kyoto University Press, 2014); Robyn M. Rodriguez, *Migrants for Export: How the Philippine State Brokers Labor to the World* (Minneapolis: University of Minnesota Press, 2010); Neferti X.M. Tadiar, "Domestic Bodies of the Philippines," *Sojourn: Journal of Social Issues in Southeast Asia* 12, no. 2 (1997): 153–191.

31 Filomeno Aguilar, *Migration Revolution: Philippine Nationhood and Class Relations in a Globalized Age* (Singapore and Kyoto: National University of Singapore Press in association with Kyoto University Press, 2014).

32 Neferti X.M. Tadiar, *Things Fall Away: Historical Experience and the Making of Globalization* (Durham and London: Duke University Press, 2009).

33 The remark was first made by U2 frontman Bono in 2003 and repeated by President Obama in 2016.

34 Renato Constantino, "The Miseducation of the Filipino," in *Vestiges of War: The Philippine-American War and the Aftermath of an Imperial Dream 1899–1999*, eds. Angel Velasco Shaw and Luis H. Francia (New York: New York University Press, 2002), 181.

35 Gonzalez quoted in Beatriz P. Lorente, *Scripts of Servitude: Language, Labor Migration and Transnational Domestic Work* (Bristol, UK: Multilingual Matters, 2017), 42.

36 Beatriz P. Lorente, *Scripts of Servitude: Language, Labor Migration and Transnational Domestic Work* (Bristol, UK: Multilingual Matters, 2017), 41.

37 In fact, it persists in expanded form, with English viewed as "minimum linguistic capital" or as the "security" language, that is wisely augmented with another language (Beatriz P. Lorente, *Scripts of Servitude: Language, Labor Migration and Transnational Domestic Work* (Bristol, UK: Multilingual Matters, 2017), 49).

38 Aihwa Ong, *Neoliberalism as Exception: Mutations in Citizenship and Sovereignty* (Durham and London: Duke University Press, 2006), 201.

39 Beatriz P. Lorente, *Scripts of Servitude: Language, Labor Migration and Transnational Domestic Work* (Bristol: Multilingual Matters, 2017).

40 Vincente Rafael, "Your Grief is Our Gossip: Overseas Filipinos and Other Spectral Presences," *Public Culture* 9 (1997): 270.

41 The issue is further complicated by the fact that a number of the family members who came from different provinces and speak another regional language preferred English because their English is far superior to their Tagalog.

3 Knocked off-script

Refusal, improvisation and disposability

with Vanessa Banta

Written criticisms of the PETA production offered by the students in a graduate dramaturgy class at the UP Diliman (in Metro Manila) made for a sobering read: "No new stories"; "White man's play for white man's problems"; "Where is the Filipino dramaturg? Sensibility of the director is not attuned to the sensibilities of the audience"; "Why are the Filipino characters speaking in English?" Despite its origins in and execution through a sustained collaboration with the PWC of BC, and co-facilitation of the PETA talkbacks by Pratt, Philippines native and University of British Columbia (UBC) PhD candidate Teilhard Paradela and Filipino-Canadian UBC PhD candidate and former member of the PWC, May Farrales, the play nonetheless was perceived to be a "white man's play for white man's problems." Further, during the PETA performances, Migrante International,[1] an alliance of Filipino migrant advocacy groups that arranged for activists and migrants to attend the production and participate in post-performance public forums, raised additional questions about audience and venue. They rightly noted that the professional production at the PETA Theater Center was largely inaccessible to many migrant workers – it was simply too expensive – and would thus do little to generate discussion among many of those whose lives are most affected by migration. Migrante invited us to more fully engage the issues by working with their organisation to bring the play into migrant-sending communities. We returned to Manila to collaborate with Migrante and Teatro Ekyumenikal from July to October 2014, a process that culminated in a community adaptation and performance of *Nanay* in Bagong Barrio – a key migrant-sending community of the urban poor in Caloocan City, the third most populated of 12 cities that constitute Metropolitan Manila.

We began our partnership with Migrante and Teatro Ekyumenikal by contributing our testimonial script as communal property to see what might evolve in an open-ended collaboration. The first step of this process involved Vanessa Banta[2] translating all but one scene from English to Tagalog. This was critical given the criticism of the PETA production and the intended audience of less privileged migrants and their families. Translating our script into Tagalog moved it more fully into the context of the Philippines and recalibrated the transcultural creative process. As a language spoken by no national peoples other than Filipinos, Neferti Tadiar has said of Tagalog that "one might call it a drop out language – simply speaking it arguably constitutes an act of defiance of the transnational."[3] Working with a Tagalog script certainly repositioned us as English-speaking Canadians, now reliant on the willingness of Filipino collaborators for ongoing informal translation, throughout the creative, research and rehearsal process.

And still whiteness clung to the play, or at least this is one possible way of understanding how we were knocked off-script as we collectively improvised towards a community performance. Collaboration, Grant Kester notes, has two meanings, one negative and another positive; this doubleness serves as a productive warning, he writes, of its ethical undecidability.[4] Working across nations, cultures, geopolitical histories, geographies of uneven development, and unequal and non-equivalent resources, the ease of collaboration could never be assumed. It was continually worked towards, questioned, and rethought. As Diana Taylor argues, the openness of performance – as lived process, praxis, episteme, mode of transmission, system of learning and embodied cognition – makes performance a particularly rich and creative site for labouring towards intercultural collaboration and solidarity.[5]

This chapter is a description of our collaboration towards a community performance in a migrant-sending community in Metro Manila. Critically, this was also an improvisation towards new research that led us to more fully understand the process of labour migration that we could not see from our vantage in Vancouver. It allowed us to heed – finally – the Vancouver dramaturg's criticism of our script: "*It's as if they [domestic workers] come alive somewhere over the Pacific Ocean.*" This criticism made note of the methodological nationalism that pervades so much migration research, and possibly also of an all too familiar capture and measure of racialised lives within the logics of whiteness.[6] It parallels Gayatri Spivak's criticism of the tendency for U.S. feminists to begin their analysis of migrant stories at the moment of landing in the destination country, and to conceive migrants' lives within the critical nexus of race-gender-class (that is, critical multiculturalism). The effect, she argues, is to reduce analyses to the binaries of black/white, poor/rich, periphery/core, to minimise critical aspects of migrants' agency and responsibility, and to miss the nuances of complex and contradictory affective tonalities and emotional registers that attend migration. Such analyses are, she argues, "narcissistic, question-begging."[7] They allow the privileged readers/audience members to remain – even if framed critically – within their own predicament of a multicultural society. A productive politics of refusal that emerged within our collaboration with Migrante and Teatro Ekyumenikal invited us to situate the lives of domestic workers coming to Canada in a more expansive geography and within attachments and relationships that extend far beyond the predicament of work in Canada.

A (partial) refusal of the script: improvising collaboration

Migrante's interest in working with us emerged out of their recognition that organising OFWs involves mobilising families and entire communities and that cultural work can be a vital means for doing this. Filipino migrant workers are difficult to organise not only because they are transient and dispersed throughout the world, but the conditions that force migration – poverty and underemployment – are all encompassing. In the words of a Migrante organiser, "with forced migration, if there's one family member who leaves for abroad and is an OFW, in the future he/she will encourage another family member. This becomes a cycle." The modest financial resources that we were able to bring to the collaboration allowed Migrante to experiment with forming a cultural wing of their organisation, which they named Sining Bulosan, to honour Carlos Bulosan, the Filipino novelist and poet who spent much of his life as a migrant in the United States. To assist in this venture, they enlisted Rommel Linatoc, a trusted member of Migrante with a long history in community theatre in Manila, and his community theatre group, Teatro Ekyumenikal, to train the community actors of Sining Bulosan, to develop the script and bring the performance into being.

Our process began with a two-day intensive workshop during which the director worked with members of Sining Bulosan, alongside the more seasoned Teatro Ekyumenikal actors. Although Migrante had hoped to recruit OFWs, they were unable to do so and the Migrante actors were youths without training as actors or, with the exception of one, direct experience as OFWs. During the initial two-day workshop, the director had the young actors improvise a variety of scenarios, which substantially broadened the scope of investigation beyond migration to Canada (Figure 3.1). There was a scene of a landlord throwing a peasant family off their land because they are unable to repay money that they had borrowed to send a family member to work abroad. In another improvised scene, a community resisted the demolition of their homes by linking arms to collectively struggle against thugs sent to violently displace them. A good portion of the workshop was spent memorising and rehearsing a long, tightly choreographed declamatory poem-tableau or *dula-tula* (drama-poem) created by the director, entitled (in translation) *Suitcase, Box, Placard.* The poem exhorts the audience to raise their placards in response to a lyrical account of a domestic worker who leaves with hope, dies under mysterious circumstances, to return home six months later, not with her suitcase but in a box, a coffin. For another scene, the actors were asked to write a letter to their mother. One of the actors, Michael, wept as he read his letter: "I hope you are not abused by your employer. I wish you were here so that you could take care of me. I am jealous of my class mates." And yet Michael had earlier revealed that he had no direct experience with OFWs. EC, who does have a history of missing intensely his mother who is an OFW, simply read, "I hope you are taking care of yourself" in quiet flat tones. The director made note of his unaffecting performance.

Figure 3.1 Rehearsals of a drama-poem.

At the end of the workshop, we reviewed the situation. The director estimated that less than half of the culminating performance would be based on the *Nanay* script. We would add the improvised scenes that bring to life the circumstances that prompt migration, new material from other migration destinations and the deadly consequences of this overseas work. We would add "Filipino performance elements." He judged the available community actors to be too young and too inexperienced for our script; the realistic testimonial monologues too dull for a community production; and Canada and the LCP too distant from the experiences of most residents of Bagong Barrio, the migrant community that Migrante had selected for the play to be performed. Though labour migration is a common experience in this community – over one-third (35.8%) of the approximately 15,000 households in Bagong Barrio depend on remittances from family members who are working overseas[8] and community organisers estimate that between 60% and 70% of households have a relative who has worked abroad at some point in time – both Migrante and the director ventured that, given the poverty of the community, few labour migrants in this community would be fortunate enough to go to Canada. Migrante's assessment is that most women migrants (who make up half of the OFWs in this community) work as domestic workers, often in difficult conditions of extreme vulnerability, in permanently temporary jobs in the Middle East or in more accessible countries in Asia.

The workshop process challenged our expectations and preconceptions. Neither the poem-tableau nor the majority of improvised scenes bore any immediate relation (we thought) to Canada's LCP and there was no mention of our script until the second day of the workshop. Moreover, the genre of the new scenes was seemingly at odds with testimonial theatre, which places emphasis on listening closely to accounts of deeply personal, real-life experiences. If there is a defining characteristic of testimonial theatre, it is that the stories told are meant to be truthful representations of what was said by a "real" person. A testimonial play provides an opportunity for individuals to – in effect – speak for themselves to an audience who is otherwise not likely to listen. It is the proximity to actual experience that is thought to confer a peculiar kind of responsibility on the audience and thereby increase the intensity of feeling and audience absorption.[9] The new scenarios developed and rehearsed during the two-day workshop were improvisations by youths who for the most part had no direct experience with labour migration, and they seemed to us at the time to be melodramatic glosses of experience. Further, the objective of verbatim theatre typically is not to provide answers but to state problems with complexity and clarity to provoke thought, analysis and discussion.[10] As discussed previously, the intent of prior performances and public forums was to disrupt existing and expected identifications and disidentifications to bring audience members into more complex relationships to the issues. And in line with Jacque Rancière's theory of the emancipatory potential of theatre,[11] we mobilised the piece to facilitate a space of unanticipated sympathies, reactions and discussion. This new production was driving towards a more singular resolution: to raise a placard and to join forces with Migrante.

And yet there are good reasons to hesitate over these initial reactions, not least because we had entered into a collaboration with an activist organisation in a new political terrain: in which a popular revolution ended martial law just 30 or so years ago, extrajudicial killings and forced disappearances attributed to the military and paramilitary continue to attract national and international attention, widely publicised political corruption undermines democratic processes, and there are underway – to this day – a number of ongoing armed struggles and distinctive peace processes. A cultural movement of poems, songs, improvised stage performances, visual exhibits and protest effigies were and remain an integral component of the resistance against numerous state repressions.[12] These works intend to convey a strong message and evoke powerful feelings with utmost urgency; they are meant to capture one and all.

Even considering popular theatre within Canada there is good reason to hesitate over our first reactions. Writing from the Canadian context, Julie Salverson questions the pre-occupation with experience and the unquestioned valorisation of testimony and documentary realism. This can disregard, she argues, "the complexity of negotiating life in the midst of loss and presumes that approaching experience as transparent maintains an innocent listening."[13] One could also challenge, especially in a context such as the Philippines, the individualism of testimonial theatre and our suspicion of collective or second-hand knowledge circulating as personal truth.[14] It is arguable that even if most of the youths in the workshop had no immediate experience as OFWs, the culture of migration in the Philippines is so pervasive that the experience is nonetheless known.

We could ask, as well, about the meaning of realistic testimonials from Canada within the Philippines, with its long history of colonialisms and anti- and decolonial struggles. Traditions of "seditious" drama have existed since the beginnings of American colonialism in efforts to decolonise the theatrical stage, sustain, recuperate and reinvent vernacular forms, and enlist theatre for progressive nationalist ends.[15] Within this tradition of political theatre, Nicanor Tiongson has argued that a Western tradition of expressionism derived from Brecht has been more popular than realism.[16] Embracing the spectacular, improvising and taking joy in performances of excess and over-the-top dramatics, and risking the play of stereotypes are other characteristics of what Lucy Burns calls *puro arte*, a Filipino performance practice that has emerged within and sometimes against a history of U.S.-Philippine colonial relations.[17] Analysing the practice of Sining Bayan in the United States, Burns argues that its use of a multidisciplinary format that combines music, acting and dance supports its objective of getting as many people on the stage as possible; this is because performances serve as valuable means to recruit new members to expand and engage political action. Sining Bulosan shares this objective.

Moreover, a barangay basketball court (where the community play was to be performed) has a different spatiality than a theatre that possibly changes which and how stories are told. It quickly became obvious, in short, that solidarity and collaboration required an embrace of multiple genres, plural objectives and that aesthetic choices and cultural work – of necessity – take shape differently in different historical, political-economic and geographical contexts.

"To refuse," writes Carol McGranahan, "can be generative and strategic, a deliberate move toward one thing, belief, practice, or community and away from another." Refusal is a concept that is "in dialogue with exchange and equality."[18] The partial refusal of our script and its focus on Canada's LCP opened an opportunity for a fuller, more equal exchange. Our first step towards this was to more fully situate the play and the issues it engages within the setting and history of Bagong Barrio and to work with the director to experiment with theatrical form. Doing new research in Bagong Barrio simultaneously brought the play closer to the intent and methodology of testimonial theatre.[19] We started visiting Bagong Barrio with Migrante organisers and Vanessa Banta (then an assistant professor in the Department of Theatre and Speech Communications at UP Diliman) to conduct interviews with long-term residents and migrants who Migrante selected for us to meet.[20] In other words, the director's partial refusal led us to both a new play and a new method of doing research through a performance. Rather than viewing performance only as a kind of research "output" or form of "dissemination" of already existing research, it became a powerful and engaging way of framing and doing research.

Both the director and Migrante were alert to the performative aspects of our white bodies and joked that our presence would attract interest and serve to promote the play within the community. Most days in Bagong Barrio began with walking together through different streets in the barrio to introduce us to one or another barangay captain in the community. These efforts were most always passively refused: the locally elected official was inevitably otherwise occupied even if a tentative visit had been arranged, but they served the purpose of establishing our and Migrante's presence in the community and made visible the organisation's network of international collaboration. At the same time, the focus of a community play led a number of residents to open their homes to us to share their stories of migration.

A renewed script

As we spent time in Bagong Barrio, we heard repeating histories and experiences of intense precarity, permanent transience and economic desperation.[21] People spoke at length about the conditions propelling overseas labour migration: the barrio's swift transformation from a community of unionised workers to a precarious labour force dependent on informal work such as peeling garlic and transforming rags into cleaning clothes. This is a process driven by the contractualisation and flexibilisation of labour beginning with the closure of surrounding industrial factories in the early 1980s and the neo-liberal restructuring of labour law. Rendered a migrant-sending community in the space of just a decade, we heard stories of long periods of family separation, histories of tremendous personal sacrifice, hard work and strategic planning, of harrowing abuse experienced in nearby countries, as well as faith in and an orientation to the future.

Against the expectations of the director and Migrante, we kept meeting the families of women living in Canada under the LCP, though Canada was clearly only one (and likely the least common) of many labour destinations.

With the director and Migrante, we committed to placing the LCP testimonies from Canada in the context of other migrant stories from this neighbourhood and, in line with the director's preferred performative genres, to use the research from Bagong Barrio to move the monologues into a more dialogical form and to write new verses that could be staged as declamatory tableaux.

And so, after Joanne, a Filipino nurse working as a domestic worker in Whistler, a recreational community outside Vancouver, tells in her testimonial monologue (substantially shortened from previous productions) of the humiliation of having to hand wash her female employer's blood-soaked panties, another woman stepped forwards in the new script to say (translated here to English, with bracketed text indicating non-verbatim additions):

MELISSA: [Blood soaked panties! Let me tell you what happened to me in Kuwait.] The 21-year-old son came into my room and stole my panties. He then masturbated on them and left them in the laundry for me to see and clean up.

[Why are we humiliated and forced to clean up employers' bodily waste all over the world?]

When Ligaya, another domestic worker in Vancouver, ends her monologue in the original production, in which she tells of leaving her children with her parents in Manila, with the triumphant: "With all of these challenges, I'm a survivor," Jocelyn – resident of Bagong Barrio – steps up to tell her story (taken verbatim from an interview) of being beaten by her grandmother and sexually assaulted by her grandfather when left by her mother in their care. The following is just a fragment of this addition to the script (translated to English, Figure 3.2).[22]

Figure 3.2 Survivors.

JOCELYN: [I'm also a survivor! I had no choice but to be a survivor because] my mother left me in the air. It is as if she suddenly dropped me from above, and no one caught me.

I was eight years old when she left.

I call myself an N.P.A. No Permanent Address. Wherever I would end up, that's where I'll be.

I was passed from one relative to the next until I stayed with my grandmother. It was in my grandmother's house I experienced being beaten, being treated like a maid by your own grandmother. They cursed at me. Then, if I wasn't able to do what they wanted me to do, they would bang my head on the wall [...] Then, the time came. It was a Tuesday night. I was working on my [school] assignment. Actually it was almost 1 am. My grandfather got up from bed and he went out of his room. Sister, would you let your grandfather who is supposedly just drinking water have his hand on your body while doing so?

The monologue of a child reunited with her mother who came to Canada through the LCP from our original play was interwoven with a verbatim monologue of a child left by a mother working in Jeddah. The following is an excerpt from the reconstructed scene, translated here to English:

CHILD OF LCP: [...] I'm laughing but I don't know what I'm laughing about, right?

ARVIN: Whenever we talk, it's like I don't want to talk to her. I would say, Yes, I'm okay. That's it. Then, she would say, 'Your aunt will get you from there.' I would go, 'Ok.' I grew up with that situation, so I'm used to it already. Our conversations were just like that.

Because my uncles told me that my mother will not come back for me anymore. Like that. It was like, in my mind I was convinced that I do not have parents. Because whenever I would have a school assignment, my aunts would not help. I was always alone. They were always at work. I had no one to talk to. The [school] would call for a meeting but they will not go. That's why I became angry. That no one was taking care of me. When they get tired of me, they would just pass me to the next family member. I grew up like that. I finally decided that I would just be the one to decide for myself.[23]

As Arvin's scene develops he tells of his efforts to build a relationship with a mother he had never lived with when he met her in Jeddah, now himself an OFW, and they decided to share an apartment (Figure 3.3):

ARVIN: Whenever we would exchange stories, we cry because we remember the past. 'Ma, if you only knew what I experienced, being passed around.' Then, she would tell me about all her experiences working there as a TNT.[24] She became a TNT because when she returned to the

Figure 3.3 Reclaiming childhood.

> Philippines once, she tried to not go back. She didn't want to return, she didn't want to renew her contract. But she said that what she earned in the Philippines will not even be enough for us. All the more in Manila, if you work here and you have to pay for rent. What will happen to my school? How would she support me? That's why she had to work hard. She was the one who worked so hard so I could go to school. She did it so she could put me through college. [....]
>
> You know, that was the first and the longest time [...]. We almost spent two years together in the same house [in Jeddah]. That's where I felt, oh this is how things are when you have parents who will take care of you. Your clothes, the time you wake up in the morning. I really felt it there, so I was very happy.
>
> Because I didn't experience it when I was young, my favorite thing we did for "bonding time" was to go to Jollibee together. We always saw children being fed with a spoon by their parents in Jollibee. I said, 'That! I want that!' I will point at the food I want to eat from Jollibee and we will eat. We will pretend that I'm still young, that I'm still a baby. I asked her, 'Ok, how would you feed me, Ma, now that we are here at Jollibee?' She showed me.
>
> I've been back in Manila for two years now. At the end of the month, I will return to Jeddah. [...] I will work as a Medical Lift Operator this time and not a driver. I think that's a safer job.[25]

These new scenes bring the LCP in direct conversation with migrant experiences elsewhere, as well as life in Bagong Barrio, in particular the lives of children left behind. This strategy is meant to disrupt the exceptionalism of the Canadian experience, such that we can ask why are Filipina migrant workers humiliated and forced to clean up employers' bodily effluent all around the world? And we can raise questions about the care of children left behind, whether their mothers or fathers are in Canada, Japan or Jeddah. By bringing the Canadian experience in close proximity to experiences elsewhere, we suggest that Canada is not the assumed "greener pasture" or "dream destination" and that migrants there share some of the same concerns as OFWs elsewhere. And as well, the same scenes take on new meaning in Bagong Barrio. When Joanne ends her monologue telling of her experiences as a domestic worker in Canada with the statement and question: "I really want to go home, but what about the fate of my family?," the questions and emotions provoked are very different when posed to their remittance-dependent family members as compared to Canadian audiences.

Reflecting his desire to create a more dialogic and dynamic performance experience, the director wrote into the now flexible, fluid and hybrid script more connective threads and reactive dialogue that allowed the performers to speak back to the Canadian experience. For instance, before and after the challenging monologue of a blatantly racist Canadian nanny agent – the only monologue delivered by a white North American in English in the Bagong Barrio performance – the director inserted the following interjections (translated here to English):

PERSON 8: They say Filipinas are calm, good with their personal hygiene, pleasant and loving, and caring. People from other nations often say this.

PERSON 9: That's what you will always hear from foreign employers. Because of our great love for our families, we would take anything for the sake of our children.

PERSON 10: My nanay [mother] will bear everything for her children even if she is treated as if her being is even lower than a dog.

[Extended Canadian Agent Monologue omitted here.[26]]

PERSON 1: You're overreacting! Isn't it natural for a people to think that their race is superior?

PERSON 4: What can we do if all that Filipinos can do is to be a domestic worker?

PERSON 3: There's nothing wrong in being a domestic worker. Because of the many domestic workers around the world, the government of the Philippines earns money!

PERSON 4: That's not what I mean, my mother is a domestic worker abroad.

PERSON 5: My sister is one too.

PERSON 4: Let me finish.

PERSON 5: OK. Finish what you want to say. This irritates me.

PERSON 4: The racist Canadian Agent is absolutely wrong.

Beyond this, the director originally envisioned that roughly 40% of the play would take shape as a traditional form of processional. His intention was for the processional to weave through the community to draw audiences into the basketball court, where the rest of the play would be performed. Though this plan was never fully executed, the idea was used to create a scene of a Tent City, calling up the situation in Jeddah in Saudi Arabia in 2013–2014, where roughly 1,000 Filipino migrants camped in front of the Philippine embassy seeking to be repatriated to avoid being arrested by a crackdown on illegal migrants by the Saudi government. The improvised scene was performed outside the basketball court prior to the main event to draw those passing by into the performance (Figure 3.4).

And finally, drawing of what we learned about the history of Bagong Barrio from long-term residents, we constructed a chorus that was used by Migrante both to advertise the play in short street skits prior to the performance and at the end of the actual performance in a tightly choreographed declamatory tableau. In English,[27]

CHORUS: We are the people of Bagong Barrio
Do you know your history? Do you know what you are capable of?
In the 1970s Bagong Barrio was known for resistance to land issues, to water issues, to housing issues
We could mobilize 7,000 people in an afternoon
We had jobs and could feed our families
And then our jobs were contractualized and our factories were moved to export processing zones
We work and we work and we work and we work and we work
We make paper bags for National Books
We sew scraps of fabric to make cleaning clothes
We peel garlic all day for 70 pesos a bag
We use our ingenuity to invent jobs out of the air

Figure 3.4 Calling the audience.

But there are no more secure jobs and people have to leave in order to survive
We need to analyze and organize
You will feel safer when you organize together
We will create our own security together
Here at home
Not in some foreign land

The chorus invoked the community's rich history of labour organising (Figure 3.5). It summoned a time from the 1970s and 1980s when Bagong Barrio was surrounded by 70 or so unionised factories in which many residents worked. Mostly Chinese- and Filipino-owned, the factories were largely dedicated to the manufacture of clothing, footwear, plastics and tires.[28] Residents active in the labour movement at the time estimate that the Bagong Barrio Labor Alliance had between 10,000 and 20,000 members in the 1970s and early 1980s, and "whenever there was a mobilization, it was easy to get 7,000 workers to come."

Substantial union organising did not, however, halt the closing of factories in the surrounding area, which began around 1983, and labour conditions in Bagong Barrio deteriorated further after the passage of the Herrera Law in 1989.[29] Enacted during the Aquino presidency (1986–1992), this law created the legal grounds for contractualisation and for the police/military to intervene in workers' strikes. Through the 1980s, many regularised workers in Bagong Barrio were ejected from their factory jobs as contractual hiring was outsourced to labour agencies.[30] As one seasoned union activist, turned migrant organiser, noted, "It was the start of what they called 'Endo' [end of contract]. After 5 months, you are out [because work becomes regularised at six months]. Then, you have to apply again." Workers who were once highly organised and unionised were thus effectively

Figure 3.5 A call to action.

transformed into a flexible and increasingly precarious labour force. It was within this context that survival became almost impossible and labour migration began on a massive scale and the memory of and faith in a capacity to intervene diminished.

Politics, precarity and performance

As a Migrante cultural event, the community performance of *Nanay* in Bagong Barrio was always already enmeshed in politics. Before the play, there were short speeches from, first, a representative of GABRIELA (one of the two GABRIELA elected members of Congress at the time), and then a councillor from Barangay 150. A third Barangay councillor also attended the play, breaking from the other councillors who, we were told that day, were collectively boycotting the event and cautioning local residents against attending our "red play." Even before the day of the performance, Migrante seemed to have met with some passive resistance: just days before the performance, they were told that the venue, a community basketball court, was available only until 6 pm (just two hours after the scheduled start of the play). Despite previously promising the facility, the barangay captain had scheduled a basketball tournament for early that evening. Only with negotiation was Migrante able to secure the facility until 7 pm. And then it became known that only half of the chairs rented from one barangay office were to be delivered.

There were numerous other unrelated challenges and setbacks on the day of the performance – seasonal torrential rains threatened to flood the entire basketball court, rented audio equipment and one community actor (who was to deliver one of the main monologues) failed to materialise. One barrio resident, who had invited some of her friends to the performance, argued that the "timing was wrong… the performance was [scheduled] at the same time as the 4Ps – conditional cash transfer for the urban poor." Others were not able to attend because of work: "They had to finish peeling the garlic by 7 pm," while concerns of heavy rain forced others to safeguard their homes in the event of flooding. And still others were hesitant to come too close to the performance, opting instead to observe from a distance; some peeping through a chain link fence, or gazing down from the balconies or roofs of surrounding houses. One of the Migrante organisers noted, "there were people who just peeped inside through the fence" or "I saw one mother was watching so far away. She could see the actors."

And yet, despite all of these complications, the play took place. And perhaps this is one of the most political aspects of the event: in conditions of extraordinary precarity, the play took place. Shannon Jackson notes that "theatre's anachronistic territoriality [which demands physical durational co-presence] might be the most interesting thing about the medium right now."[31] The temporary occupation of public space and the physical, bodily assertion of principles of equality through public assembly in the midst of precarity is highly significant. In Judith Butler's phrasing, "When the bodies of those deemed 'disposable' assemble in public view, they are saying, 'We have not slipped quietly into the shadows of public life; we have not become the glaring absence that structures your public life.'"[32] They are "exercising a plural and performative right to appear, one that asserts and instates the body in the midst of the political field."[33] In the context of a neo-liberal ethos that individualises responsibility and celebrates the entrepreneurial self, public assembly is a disruptive "assertion of plural existence."[34] Theatrical performance almost always involves sustained collaboration through time and space[35]; in circumstances as precarious as those of Sining Bulosan, this was an accomplishment in and of itself.

Although it is challenging to gauge community reaction and good reasons not to over claim its effects,[36] we know that for some the performance in Bagong Barrio was deeply felt. From a Migrante organiser we heard that one resident, herself a former OFW, was "just speechless, she just cried," because she saw the experience of OFWs in other countries. From an older Migrante member who had made the promotional posters,

> It was easy for me to cry. I didn't know that was the story. I wasn't there for the rehearsals. The first time I saw it, I cried. In the beginning [after the show] I didn't want to approach anyone. I felt very emotional. I remember my siblings in Jordan – they were there for ten years. We didn't know what the conditions are. We don't know if they are raped or exploited. It made me angry because it was clear from the stories that the government is not providing support and not stopping migrants from [being exploited/abused]. So now it is clearer to me.

As a talented illustrator, he was intent on turning the play script into comic book format.

Speaking with a long-term resident, Marilou, two days after the performance, she reported that those who attended were "excited that we have a show like that." The children, in particular, (about 30 children sat on tarps on the ground at the side of the performance space) she said, "understood the topic… They weren't bored by it. They were very attentive. All the children." Finding one child crying after the play, she asked him: "'Why are you crying?' [He said:] 'I miss my mother.' A nine-year-old boy! Christian is his name. He was left with his grandmother since birth [his mother is working in Canada in the LCP]. But now, because life is hard, [his grandmother also] works as a domestic helper. So, his tita [aunt] is taking care of him now. He's not happy. Of course, he feels it that his mother is gone. Our children were able to relate to the play." A Migrante organiser told of meeting another resident after the play who said that her daughter also had to work in Canada. Her daughter was trained as a nurse, but was working as a caregiver: "Her mother was crying when she was telling me the story. Because the children often tell their grandmother they miss their mother." Marilou also told us this about the community reaction to the play: "They were curious. 'Imagine, the writer is a foreigner.' But, I told them: 'That's our story, based on the [interviews done here]. It's our lives.'"

As important and perhaps more enduring are the effects felt by Sining Bulosan youths. Given all of the setbacks that plagued the performance, Migrante had met with Teatro Ekyumenikal just the week before the scheduled performance to consider cancelling the show; it was the young Sining Bulosan actors who insisted that they be allowed to persist. In the assessment the day following the performance, the capacity to work through all of the challenges – including the disappearance of the director due to ill health and other commitments, along with several of the original Sining Bulosan members – was judged to be one of the main accomplishments. From the Migrante organiser in charge of the production, "I can see the changes since we started: not just in skills but in people. You stood your ground. You did not give up, despite personal problems. As a collective we helped each other to overcome our struggles. That's why we are still here." Even those who could not continue remained, he told the young Sining Bulosan members at the assessment, because some of their stories are now embedded in the script. He noted that this process began with their performances of the Bagong Barrio verse as a short street play: "The[se performances] had another purpose. They led to breakthroughs and new revelations. They weren't just to advertise the play. They helped to build confidence and shed fear. They are a way of talking to people. They are important because that skill is important for building the company." In effect, they were a way of learning to be an organiser.

For some, one significant aspect of the play was that this was the first time they had heard the history of Bagong Barrio, its substantive history of union organising, of strikes and protests during the turbulent repressive years of martial law (1972–1981), and the more recent experiences of OFWs in the community. Paul, one of the community actors, said, "This is my first time in a theatrical production [...] One part I really like are the struggles of the people in Bagong Barrio: their fight for their homes, water, electricity. I am happy that it was adapted to the experience of Bagong Barrio. It made it come alive. The production helped in getting us all here, binding us together. I was surprised that there were audience members who cried because they saw their experiences." Or as one Migrante organiser and resident of the barrio noted,

> I saw the history of Bagong Barrio for the first time. Before I only heard of it from other people but had not understood it that way. That's why I think that theatre production is a powerful tool for understanding stories. The interviews that we did, those interviews helped us to understand what Bagong Barrio is about [...] I'm happy that we did the play even if in the beginning I didn't give it a chance. We did it. In Bagong Barrio we are open to the possibility of continuing the project and we hope that we are able to take it to other areas.

In a post-performance assessment, Migrante judged the project to be a success; they had succeeded despite many personal and organisational challenges, recruited new members through the process, and the performance contributed to building confidence and organising capacity among youths in Bagong.

If claiming public space is in itself a politically significant event, the transience and persistence of our one-off, precarious community performance is additionally significant because it signals that "it could happen again."[37] As indeed it has. Sining Bulosan has performed scenes from the play at mass demonstrations and at other Migrante events. At the performance in Bagong Barrio, a Migrante leader was approached by a priest from the neighbouring barrio of Sta. Quiteria, who asked about the costs of restaging the play in their parish as part of their Migrant Ministry programme. This initiated a relationship between Migrante and members of this parish church. Because the majority of the actors in the Bagong Barrio performance were from Teatro Ekyumenikal (whose commitment to work with Sining Bulosan had ended), Migrante asked that the parish church involve their theatre group as well, and rehearsals opened an opportunity for Migrante to educate these community actors about migrant conditions. The script was further adapted to the circumstances. From the Migrante member in charge of the production,

> Because the end of that [Bagong Barrio] performance was focused on Bagong Barrio or on the stories of the individuals, I had to create one chorus that discusses the collective experience of all so that we could use the script in other places too. We talked about it and there was a collective decision made regarding these changes. But, similar to other materials for community theater, the piece remains open to any change depending on the length of time and process of creating or mounting the production. If the theater artists are able to immerse or do more research in the places where they will stage the performance, they can adapt it and the piece will speak more to the specific experiences and feelings of the audience of the play.

The play was performed in Sta. Quiteria on November 28, 2014, as matinee and evening shows that attracted an audience of roughly 300.

As for the future of the play in Manila, it is unclear and evolving. In summer 2015, a continuing member of Sining Bulosan asked the former producer for the script. Different segments of the script have been performed in different contexts. On the other hand, the script may fall away and Sining Bulosan may work more fully with dance and rap. But we were told that "Migrante recognizes the potential of the project in all aspects – organizing, developing leaders and even as an income generating project (IGP)." The former producer, who has been moved to another region to organise migrants there, holds on to the prospect of working with theatre: "I'd like to try what our cultural activists in the '70s did. They had material they always used in rehearsal. This was their way to recruit new members, while they still responded to the invitations from other organizations to perform for their campaigns, in-door activities etc. This way, they would always meet and the cultural workers were always active." This vision places our collaborative script, ever changing and being adapted to new circumstances by new members, at the disposal of these cultural workers, possibly exemplifying the "rhizomatic potential of interculturalism – its ability to make multiple connections and disconnections between cultural spaces – and to create representations that are unbounded and open, and potentially resistant to imperialist forms of closure."[38]

Returning with an archive

In what is now a highly influential distinction, Diana Taylor contrasts the archive, as authorised place for storing objects and a system of classification, from repertoire (or performance).[39] As noted in the introductory chapter, the archive sustains a form of knowledge that is rational, linear, individualised, enduring, and works across distances at a remove from the knower. Repertoire, in contrast, is embodied, ephemeral, requires presence and is a kind of "knowing in place."

As we worked with Migrante to situate our play in Bagong Barrio, we assembled new materials that embed the LCP in a much more expansive analysis. We documented the production by the Philippine state (in the context of a history of collaboration with the International Monetary Fund and World Bank to attract foreign capital) of large portions of the population as surplus, the value of which lies (at least to the state) in the remittances OFWs send home and the reduction of unemployed populations in the Philippines. Beyond improvising day-to-day survival, the research that we conducted with Migrante also documents how OFWs purposive investments sustain the production of OFWs over generations. Dedicated investments in children's education, in many cases, buys their children privileged access to overseas employment and serves – for better or worse – to (re)produce the next generation of labour exports. This system of renewal meets the needs of wealthier countries such as Canada for well-educated, low-paid flexible workers, allowing these wealthier countries to offshore the costs of social reproduction. In a cruel twist of fate, these highly educated workers are often deskilled in the process of migration,[40] or returned home depleted after lifetimes of work in another country. OFWs in Bagong Barrio also make strategic investments in housing, which – like education – is an unstable economic investment and asset, largely because many households have uncertain claims to land title. The threat of urban displacement is ever present within the highly inflationary Metro Manila land market.[41]

Beyond hoping and anticipating that these research materials will circulate as repertoire in ways that are useful to Migrante, we make use of the authority of the scholarly archive here to record some of the stories told to us in Bagong Barrio. We archive two further constructed scenes from Bagong Barrio, taken directly from verbatim testimony (with editorial additions marked by brackets). We do this because these are stories of lives too easily and simply rendered as mass, surplus and disposable, to attend to the specificity of the narratives of those otherwise conceived as disposable within surplus populations.[42] This disposability is compounded by the fact that they were told in Manila, a context that itself is often consigned to the margins of Western scholarship.[43] Aside from its location in the global south, Manila's trauma of martial law, the fact that 80% of the urban infrastructure was destroyed in the Second World War, and over one-third of the residents still live with precarious rights to housing in informal settlements[44] are added characteristics that make it a city that is especially susceptible to erasure: a "history of disappearing and forgetting" is embedded, Neferti Tadiar argues, "in [Manila as an] urban space."[45] In this context, the enduring, distance-defying quality of the archive and a realist testimonial genre that attends with care to the details of individual lives seem to hold radical potential.

These are important stories to tell as well because some are excessive both to a narrative of disposability and to the migration narrative so prevalent in the Philippines. Following Tadiar, paying close attention to detailed life stories can open opportunities to listen beyond established scripts, for what "falls away" from these accounts, for overlooked modes of social experience and social co-operation that elude our existing narratives. Attending to what falls away, Tadiar writes, "opens up the possibility of other genealogies for understanding those remaindered ways of *living* in the world that move and generate that world in ways we would otherwise be unable to take into political account."[46] We have tried to listen closely to what residents of Bagong Barrio told us about their lives, for ways of living that are both subsumed within and somewhat excessive to accounts that might render their lives as merely waste or wasted within global capitalism. We share these life narratives – constructed as scenes, not with the intent of stabilising and reclaiming meaning through the archive, but to put life stories from Bagong Barrio into wider circulation, moving towards unknown and unanticipated destinations. The first conveys some of what we learned about the intergenerational production and wearing down of migrant workers, the second of desires beyond labour migration.

Scene one: cruel optimism

LUIS: We were living hand to mouth. So, I told myself I will continue my studies. My wife was selling items so she supported me somehow. She said, 'okay, study'. By the mercy of God, I finished vocational school. That was when I had the courage to [migrate]. That was in 1984. I left June or August. Even if I didn't want to, thinking of your family it doesn't matter if you want it or not. Expenses were increasing, your salary decreasing.

When I went abroad, I went to Saudi Arabia. During my first months, sadness really defeated me. It was in Saudi where I learned the saying of workers, "Welcome to this prison without bars." My first run, I was really defeated by sadness. I was there for 14 months only. When I left Saudi, I said, I don't want it anymore. I don't want to return to you.

The second time, I returned and applied for a maintenance job at a hospital of the Ministry of National Defense. I worked there for over 20 years.

I went home four times in those years. [All I could think about was] money. As long as I send what they need here it doesn't matter so much if I come [home]. [Because] whenever you go home, the first thing you think of are the gifts you need to bring. [And] when you get here and you see your family, you want to treat them by taking them out. What if you don't have the money to do so? Where is your heart? Or is it love too when you sacrifice going home as long as your child holds chocolates? Even if you are not there, you are happy [knowing this]. Am I right? [If] you go home you carry the burden of spending for everything. In [all] my years abroad, I went home four times. I don't regret it. Because, why? You will go home and there are people and things you love – yes. But even that causes you great pain.

My experience in [Saudi]? I worked hard even though my salary was not high and not low. It was just enough. Hand to mouth. I was sending money and she was working. Even if she was working and I was sending money, we didn't have enough.

[Joyce enters and faces the audience directly.]

JOYCE: [Having him overseas:] It was nothing, it was natural. Of course, you have to help out your husband. You have to raise your children, for him, for the whole family. [But] it's hard to be a mother and a father. It's hard because you are in charge of everything. Your children, your house. It was very hard…

When my husband went abroad, the money he sent was able to help but it was also not that big. I had to budget it in the right way. Everything was budgeted. All the money I got I stapled and divided. The money to pay for the electricity, water, tuition, transportation, allowance, school projects. You couldn't do anything beyond that. You couldn't subtract anything from that. And if I felt we were about to run out of money, I knew it was time for me to work harder.

I had so many jobs. imagine in a day: in the morning I sold dried fish. Then, I would do the laundry for someone. At 3 pm, I would start selling again. At night – you know that Shell gasoline station over there? – I will sell Balut.

My husband was away for 28 years, I didn't even experience rewarding myself or relaxing. That SM [shopping mall] over there? I couldn't enter that. I would enter it when he would go home and he would say, 'Let's go out.' I told him, I teased him, 'That SM, I will get lost in it.' How would you enter it when all your money is already stapled? You couldn't unwind. I couldn't even go to the movies. I only watched once. I was forced by my friends. They said, 'Come!' Even if I didn't want to, but because I was teased [...] I only experienced it once. That was not repeated.

To tell you the truth, if you don't know how to budget your money, nothing will happen for you. This house, I was able to build [piece by piece]. First, I bought hollow blocks. Until I was able to collect 650 hollow blocks. What came next were the steel pipes. That beam, I asked what do I need to build that? I would just buy! I would buy the materials right away [...].

If we didn't do that, we would just have our mouths open. You know about money, right? It goes away fast. That's why I told him, let's [build this house] so that we have another source of income, that's sure and reliable, right?

[With all of that I managed to put our three daughters through university.] He came home for the graduation of my youngest from high school. The next time was when the youngest graduated from college. There were cell phones then already. I remember [telling] him [on the phone at the other daughters' graduations], "Her name is about to be called! I will let you listen." I told him, "When her name gets called you will hear it." I always cry when I remember. "Listen, your child is being called!" We did that until all of them finished.

LUIS: Saudi made me go home. It was part of the implementation of Saudi Arabia's policy. I don't want to brag about myself but they really didn't want me to go. They still needed my service but the time came when a royal decree stated that after 60 years [you have to retire and return home].

Still, I was happy to go home. [But] now, there are no jobs for me here. I'm thinking of leaving again. Honestly, here in our country [...] even if you have a lot of experience, nothing. That's what I notice. Like us OFWs who could still work here. They say there's an age limit.

Really your body can still work. But we can't do anything because there are so many Filipinos who are unemployed. I read the news and I know that there are many others who are looking for jobs. I don't try to apply here anymore because I know no one will accept me. Also, my wife wouldn't let me.

JOYCE: He has many illnesses.

LUIS: I think that's what happens when you really do not have anything else to do. Your body will desire to work because you got used to. That's what I would consider my experience with work.

JOYCE: [Me,] I have a sidecar. I pick students up from their homes then take them to school. Yes! I'm the one who drives it! My husband and I were fighting when he said 'Who will drive that?' [hoping that he might.] I said, I'm in charge of that. When classes opened, I went to the school. I asked: 'Mothers, who wants this service for their kids?' They said, 'Me, me, me!' There were 20 of them! …

My eldest daughter is in Saudi. She's a nurse. Then my other one, she went abroad too. She studied computer science but she took [certification for] CSSD: sterilization of all medical instruments in the hospital. But she only stayed there for a year. She encountered a problem there.

She got pregnant. Her boyfriend committed suicide. He hung himself. If you're pregnant there, of course, that's illegal, right? It's going to be a big problem if she gets stuck there alone. So, my other daughter helped out. She said, 'don't say anything.' I don't know what they did. All I know was that my daughter said, 'I got it, ma.' They were able to get her out of there. She gave birth here.

Of course, she cried and cried. Even her sister was crying. Me? As much as possible, you won't see me crying. I would fight. I would say, 'What's the problem? Let's figure out a way!' You have to be strong when faced with a problem. You have to be wide and open-minded when it comes to a problem. Assess, what is the bigger problem? You go there first. Find a solution. We have to be strong when it comes to problems.

Scene two: what falls away

Tala is an energetic, charismatic 17-year-old girl with a slightly mischievous "tomboy" demeanour, stylish in t-shirt and cargo shorts. She has short hair. She holds an iPhone, which is always prominent as she talks.

TALA: My cousin was working here in bars as a GRO. [Guest Relations Office: she gets customers to buy her drinks.] Of course she was not earning enough and she can't work that way forever. She dared to leave the country because she couldn't provide. She went to Dubai. No one knows what her job is. We were all surprised then when she called to tell us she's in jail.

I think that was last year. When she called, I was about to ask her to send me something, 'Ate, buy me a t-shirt.' Then, my aunt said, 'How would she buy that for you if she's in jail. Hala! She's in jail.' My aunt said that her visa and papers were confiscated and hidden. Of course, if you get caught doing something and you don't have your documents, you will be put to jail.

My aunt couldn't get a hold of her, and my cousin hadn't called for months. And then, when they finally got to talk, that was the time my cousin was just released. She was crying, she wanted to come home. But, of course, because she couldn't find a good job here, she went back to Dubai to take a chance again. She was here for a week, then she was gone.

Of course, if it's your relative, you really worry. However, everyone who lives here is poor and having a difficult time. They can only focus on one thing: work. My father, all he does is work. [He drives a jeepney and sometimes works as a mechanic.] Then, my mother peels garlic. Of course we don't have enough.

In Bagong Barrio, all our houses are connected. There are six families [in my mother's extended family, all living side by side in the same building]. [We've lived here almost all my life.] My father always says my mother's navel is buried here. She doesn't want to leave. We have lived in other places, but we always end up here. We always come back to Bagong Barrio.

We return because, like, when we were in Pampanga, we had our own house, but we didn't have a source of income. It was in Pampanga that I experienced [...]. You know when you go to the river and you try to find, "Here! Here's a piece of metal!" I was only seven. We were scrap collectors. My mother would do the laundry for others. Then, sometimes we children worked as load carriers at the market. We would carry melons and get a melon as payment.

Whenever I think about that time now, I think that's such a hard thing to do – it's like when you're going to eat, you think I better eat a lot because there's a chance that I won't be able to eat the next time.

We came home to Bagong Barrio when I was 10 or 9 and we've never left since. My grandmother was able to get a big piece of land [when she first squatted here in the 1960s]. We're happy here. We are happy we live in Bagong Barrio. But there are times when we don't eat because we don't have anything.

I want to improve our situation. That's my first dream. I want to build us a house that's just for us because that place we have right now, [the rumour is that it] will soon be demolished. I like imagining things. I would buy a house just for my family, a house this high, or with this color. Everything I imagine! I want my house to have at least four floors. I want it high. Then, on the first floor, we have a small store or a grocery. Anything as long as it's some sort of business to help my family. Here in Bagong Barrio. That's where I grew up. I don't want us to live in an executive village [in Manila]. I grew up like this; why would I want to go to a village? I still want to be [here]. My family is always included in the plan. I want their lives to be better when my life is better. I want us all to be successful together.

Do I want to go abroad? I used to. When I was about 15 years old, I wanted to be on an airplane. I wanted to work in a different country because I thought I would earn more. Now I want to go to places like those if I am doing better so I could just experience them. I want to go to Paris. I want to go the Eiffel Tower and take a selfie.

I want to live [here] because it's where my lola grew up, grew old and died. I really just want to stay here. If the president of [National Housing Association] dies, maybe they will forget [to] demolish [our houses]! [When they come to demolish] I would protect our place. But can I do that? Can I go against them? I think the demotion team will lose. Other

> residents in Bagong Barrio have guns. When their houses get demolished, pak! People would drop dead in their camp. It's possible that everyone dies. But they won't be able to demolish us completely.

The first scene is a composite from the testimony from two long-term resident households of Bagong Barrio, in which both men worked in Saudi hospitals for over 25 years for wages that could not support their families, and who were forced to return at age 60. While it possibly exemplifies what Anishinaabe scholar Gerald Vizenor conceives as survivance: resistance ("you won't see me crying") through survival and resilience,[47] it also archives the repetition of precarity across generations despite doing everything an entrepreneurial subject can do: invest in education and the cultivation of self, work hard, manage resources carefully and strategically, take calculated risks and invest wisely in income-generating opportunities. It enriches the script that we took to the Philippines immeasurably because it embeds the largesse of countries that offer temporary work opportunities (the proverbial win-win-win situation for sending and receiving countries and migrant workers alike[48]) within the stark reality that many smart, educated, articulate, diligent, hardworking and resourceful migrants barely manage to survive and the costs of social reproduction are absorbed by families at home. Documenting the intergenerational reach of precarity raises the ethical and political stakes. As Elizabeth Povinelli argues, sacrificial redemption is a technique of social tense that is often deployed to explain (away) social harm.[49] The degradations of temporary labour migration are often justified for the benefits that come in the future. A close appraisal of the repetition of disposability across generations puts a lie to this mode of deflection, used in the global north and south alike to legitimate labour migration, because it makes clear that sacrifice in the past and present is unlikely to be redeemed in the future.

Tala tells a very different story. She is deeply rooted in Bagong Barrio. While her cousin worked as an OFW in Dubai, and was jailed there, her family's history of migration, propelled by the same precarity that leads so many Filipinos to work abroad, is internal to the Philippines. Her grandmother was among the first residents of Bagong Barrio in the late 1960s, and her father says that her "mother's navel is buried here. She doesn't want to leave." Desperation led her family to leave on one occasion when Tala was seven years old. They moved to Pampanga, where the children collected scrap metal and plastics from the river and loaded trucks at the market to supplement the family's meagre income. The interview with Tala, from which this scene was constructed, was surprising because it revealed an aspirational geography unfamiliar to Canadians. The provinces hold no allure to her: they are the site of her life as a garbage picker and a scene of hunger. Tala also refuses to inhabit the cruel optimism[50] of the fantasy of upward mobility when she scoffs at the idea of living in an executive village. She refuses to access her rights to "migrant citizenship"[51] when she declines a future as a migrant worker. Her interest in travelling is touristic: Canada and the United States would be nice to visit, and she desires Paris as a site for a "selfie" and not a job as a migrant. She wants to live in what on the face of it seems an unloveable place, Bagong Barrio, because this is where her lola grew up, grew old and died. This is where she and her family have lived most of their lives, happy but for the lack of food. She aspires to buy a big house for her family in the barrio, with space at street level for a business that will sustain her family. What is excessive to conditions of disposability is her love for her lola, her concern for her father's health and her tenacious hold on her family's history in Bagong Barrio. She can imagine – "Pak!" – the defence of her space: "They won't demolish us completely."

Her family is central to her aspirations, and her desire is for them to succeed together in the same place. Narratives of disposability and overseas migration do not absorb her aspirations, her insistent life-making and her fierce claims to place. Her refusal to subject herself to the kind of self-lending that Tadiar sees in domestic workers might be another, albeit very different way of thinking about what falls away, at least from the perspective of Canada, where it is presumed that Canada always already is the desired destination for Filipinos if given the chance. So too, Dylan Rodriguez writes of the "arrested raciality" of Filipinos within the Philippines and the mass internalisation of the dream of migration to the United States. The peculiarity of American colonialism in the Philippines has fashioned the United States, he argues, as "a site of redemptive and existential progress," and as an object of almost religious desire.[52] The radicality of Tala's words is that she eschews that dream and simply wants to stay put in a troubled barrio and in a city known for its pollution, noise, congestion, poverty and systemic corruption.[53] This might be one small but significant gesture towards displacing what Rodriguez identifies as the global "structuring dominance of white life."[54] The same testimony takes on different meanings in different contexts. In Canada, Tala's words profoundly unsettle Canadian immigration policy insofar as the poverty of the Philippines is routinely called up to justify almost any degradation under the LCP, on the supposition that in comparison to the Philippines the possibility of migrating to Canada under any conditions is worth it. In the Philippines, they strike at the heart of the so-called "colonial mentality." Her testimony, in other words, in different ways, challenges both ends of the global care chain. We archive Tala's words with the hope of bringing different audiences close enough to feel their import.

Collaboration and repertoires of learning and unlearning

The collaboration with the PWC of BC began in participatory research, and feminist, anti-racist scholarship; collaborating with Migrante in the Philippines opened new questions about insidious and enduring patterns of colonialism and what solidarity towards decolonial futures might mean.

We are cognisant of criticisms of the drift of decolonising language into approaches that might previously have been described as social justice or anti-racist perspectives[55] in ways that weaken the specificity of claims of decolonisation (to land and other resources) and recentre whiteness. Without claiming too much, we return to the messiness of collaboration through performance as a site of learning (and unlearning), negotiation, shared labour and "the mindful surrender of agency."[56]

Following Denise Ferreira da Silva's distinction between the "transparent I" of the white European subject and the "affectable I" of the racialised mind/body subjected to laws of nature and the supposed superior force of Europeans,[57] Dylan Rodriguez suggests that perhaps the most promising strategy is not to claim some kind of authentic radical possibility for the Filipino subject/body but to undo the presumption of the "social and philosophical coherence" of the transparent white subject and to displace its "presumed dominion over the lived meanings of race/place/body."[58] In a small way, this is perhaps part of the process in which we were engaged with Migrante. The LCP came into focus differently in Bagong Barrio in ways that opened up new perspectives on it, and it is only part of the world of migration viewed from Bagong Barrio. We met families in which one daughter worked in the LCP, another as a "seafarer" as a physiotherapist on a cruise ship, another on a farm north of Manila. A child of the LCP cried at the play not only because his mother was in Canada, but because the precarity of his *lola* had forced his grandmother to work as a domestic helper in Manila. Some aspects of the play for which we had little sympathy grabbed the attention of and moved audiences to tears. The confusion of genres was only confusing to us. We unlearned some of what we thought we knew, and the play – in its travels – opened our analysis in unexpected ways.

Notes

1 For more on Migrante's organising, see https://migranteinternational.org.

2 The research for this performance was done in collaboration with Vanessa Banta, then faculty at UP Diliman, now a PhD student in geography at UBC. For some of her own research/performance/pedagogy, see Vanessa Banta, "Empathetic Projections: Performance and Countermapping of *Sitio* San Roque, Quezon City, and University of the Philippines," *GeoHumanities* 3, no. 2 (2017): 328–350.

3 Neferti X.M. Tadiar, *Things Fall Away: Historical Experience and the Making of Globalization* (Durham and London: Duke University Press, 2009), 171.

4 Grant Kester, *The One and the Many: Contemporary Collaborative Art in a Global Context* (Durham and London: Duke University Press, 2011).

5 Diana Taylor, *The Archive and the Repertoire: Performing Cultural Memory in the Americas* (Durham and London: Duke University Press, 2003); see also Helen Gilbert and Jacqueline Lo, *Performance and Cosmopolitics: Cross-Cultural Transactions in Australasia* (Basingstoke: Palgrave Macmillan, 2007); Ric Knowles, *Theatre and Interculturalism* (New York: Palgrave Macmillan, 2010).

6 Katherine McKittrick, "Mathematics Black Life," *The Black Scholar: Journal of Black Studies and Research* 44, no. 2 (2014): 16–28.

7 Gayatri Chakrovorty Spivak, "Thinking Cultural Questions in 'Pure' Literary Terms," in *Without Guarantees: In Honour of Stuart Hall*, eds. Paul Gilroy, Lawrence Grossberg, and Angela McRobbie (London and New York: Verso, 2000), 335.

8 Migrante, "Community Profile," n.p., n.d.

9 Robin Soans, "Robin Soans," in *Verbatim Verbatim: Contemporary Documentary Theatre*, eds. Will Hammond and Don Steward (London: Oberon Books, 2008).

10 Will Hammond and Dan Steward, eds., *Verbatim Verbatim: Contemporary Documentary Theatre* (London: Oberon, 2008).

11 Jacques Rancière, *The Politics of Aesthetics: The Distribution of the Sensible*, trans. Gabriel Rockhill (London: Continuum, 2004).

12 Alice Guillermo, *Protest/Revolutionary Art in the Philippines, 1970–1990* (Quezon City: University of the Philippines Press, 2001).

13 Julie Salverson, "Change on Whose Terms? Testimony and an Erotics of Inquiry," *Theater* 31, no. 3 (2001): 121.

14 As a case in point, some of the controversy that surrounded the testimony of Rigoberta Menchú turned on the fact that she reported others' experiences as her own and narrated second-hand reports as first-hand experience. Mary Louise Pratt noted in relation to the controversy that the category of personal narrative works differently in Western metropolitan and non-Western indigenous cultures. See Mary Louise Pratt, "I, Rigoberta Menchú and the 'Culture Wars,'" in *The Rigoberta Menchú Controversy*, ed. Arturo Arias, 29–48 (Minnesota: University of Minnesota Press, 2001).

15 See Lucy Mae San Pablo Burns, *Puro Arte: Filipinos on the Stages of Empire* (New York and London: New York University Press, 2012); Doreen Fernandez, *Palabas: Essays on Philippine Theatre History* (Manila: Ateneo University Press, 1996); Neferti X.M. Tadiar, *Things Fall Away: Historical Experience and the Making of Globalization* (Durham and London: Duke University Press, 2009); Nicanor G. Tiongson, *Dulaan: An Essay on Philippine Theatre* (Manila: Cultural Center of the Philippines, 1989).

16 Nicanor G. Tiongson, *Dulaan: An Essay on Philippine Theatre* (Manila: Cultural Center of the Philippines, 1989).

17 Lucy Mae San Pablo Burns, *Puro Arte: Filipinos on the Stages of Empire* (New York and London: New York University Press, 2012).

18 Carol McGrahahan, "Theorizing Refusal: An Introduction," *Cultural Anthropology* 31, no. 3 (2016): 319.

19 Geraldine Pratt, Caleb Johnston, and Vanessa Banta, "Lifetimes of Disposability and Surplus Entrepreneurs in Bagong Barrio, Manila," *Antipode* 49, no. 1 (2017): 169–192.

20 As researchers, we were positioned differently. Johnston and Pratt are white, Anglo academics while Banta is a Filipino native of Manila with a history of participation in migrant organising. Some interviews were conducted entirely in Tagalog and later translated, others in a mixture of English and Tagalog. Those we interviewed were variably involved with Migrante: some fully committed, others finding their way to the organisation, and others associated only tangentially. In some cases, Migrante organisers were hearing life-stories for the first time and thus the interviews were a research opportunity for their organisation as well. Most interviews were conducted for the purpose of developing the community adaptation of *Nanay*, and residents understood that their testimony would inform the play and likely be integrated into it. In this sense, they understood themselves to be testifying to their community as much as being interviewed by us. We cannot make claims about the representativeness of those with whom we spoke. Migrante made an effort to introduce us to different kinds

of community members: to three older residents with deep roots in the community, either in organising or in the community-based financial cooperative, and to five OFW families. These informants were able to speak about others on their street and in the neighbourhood. We also interviewed Migrante organisers who generously shared their own research and experience in the neighbourhood, and five youths from the area who were involved in the play and newly involved with Migrante. We were able to return to interview some of the informants a number of times and to attend Migrante forums in the community.

21 Bagong Barrio was settled by informal settlers during the 1960s, by migrants moving into the metropolitan area from poor rural areas from northern and other provinces, and by those displaced from extra-legal settlements located in the Intramuros and Sampaloc areas in central Manila. After Philippine President Ferdinand Marcos established the National Housing Authority in 1975, an infrastructure of roads, sewage and electricity slowly developed. Succeeding presidents have persisted with the promise (and failure) of delivering land-titling to residents, most recently by President Arroyo in 2005 during her "de Soto tour" of poor communities in Metro Manila. Many Barrio residents thus remain vulnerable to eviction and displacement within Manila's aggressive urban restructuring and extraordinary land inflation.

22 Nagpapasalamat ako na tinuro sa akin ng nanay ko ang maging matapang. Matapos lahat ng mga ganitong pagsubok, isa akong survivor!

Jocelyn: Survivor din ako! Napilitan akong maging survivor dahil iniwan ako ng nanay ko sa ere. Bigla akong bumagsak at walang sumalo.

Eight years old po ako noon, nung umalis siya.

NPA po ako. Non Permanent Address. Kung saan lang ako mapunta, dun ako.

Pinagpapasapasahan ako hanggang napunta ako sa lola at lolo ko.

Doon ko naranasan ang pambubugbog sa kamay ng kapamilya. Ginagawa akong katulong ng sariling kong lola. Minumura ako. Tapos pagdi ko nagawa ng tama yung pinapagawa sa akin, iniuuntog ako sa pader.

Since grade four hanggang six working student ako. Kahit nung first year hanggang third year, working student pa rin.

Trabaho sa umaga aral sa gabi. Naging baby sitter po ako ng isang abogado. Ang sweldo ko lang noon ay 1,500 a month. Tatlo yung bata. Tatlo rin naman po kaming katulong pero dalawang bata ang inaalagan ko. Syempre pagdating ko na galing ng school, bibihisan mo yung mga bata. Di ka na makakatulog kasi papasok din yung mga bata. Aayusan mo sila, sasamahan mo sila sa school. Minsan meron pa silang activity na magaral sila ng taekwondo, ballet, voice lesson. Kahit mahirap, tiis.

Tapos nagkita kami uli ng lola ko, 16 ako nun. Hindi sana ako sasama kung hindi lang sinabi sa mga amo ko na idedemanda daw niya as child abuse. Kasi nga yung age ko. Sabi ko, dito na kasi ako halos parang lumaki sa amo ko eh. Pero siyempre para hindi magulo, sumama ako.

Akala ko nagbago na siya. Akala ko tutulungan na niya akong makatapos. Noong unang dating ko dun, oo, ambait. Maasikaso. Siguro wala pa sa isang linggo, bumalik siya sa ganoon. Ginagawa niya uli akong katulong.

Tapos dumating yung time na. Martes ng gabi yun, gumagawa ako ng assignment. Actually magwa-1 o'clock ng umaga na noon, tapos bumangon ang lolo ko, lumabas siya sa kwarto. Ikaw ba ate papayag ka na iinom lang ng tubig yung lolo mo, yung kamay niya nakahawak pa sa katawan mo?

23 **Child of LCP:** Tumatawa ako pero di ko alam kung ano ang nakakatawa.

Arvin: Pag naguusap nga kami ng mama, parang ayaw ko na nga siyang kausapin eh. Oo, okay lang ako. Yung ganon lang. tapos, sasabihin lang na, kukunin ka na naman ng tita mo diyan. O sige, yan na naman kasi yung kinalakhan ko na eh. Ganun lang.

Sinasabi din ng mga tito ka na yung nanay mo di ka na uuwian. Ganoon. Yung parang, nag- ano na yung utak ko, na wala na akong magulang. Kasi kung tuwing may assignment ako sa school, di naman ako natutulungan ng tita ko. Lagi akong nagiisa. Nasa trabaho sila. Parang nagkaroon ako ng galit. Kaya sabi ko ako na lang ang magdedesisyon para sa sarili ko.

24 Tago Nang Tago, undocumented or "in perpetual hiding."

25 Kapag nagkwekwentuhan kami ng mama ko, nagkakaiyakan kami, na naalala yung nakaraan. Ma, kung alam mo lang yung naranasan ko na pinagpapasahan ako. Tapos, nagkwento din siya kung ano yung karanasan niya sa Jeddah bilang TNT. Nag TNT kasi ang mama ko kasi, noong umuwi ang mama ko sa Pilipinas isang beses, sinubukan niya na sana hindi na bumalik sa Jeddah. Pero hindi daw talaga sasapat, sa amin ang sahod na kikitain niya a Maynila lalo na't mangungupahan pa kami. Eh, papaano yung pagaaral ko? Paano niya masusuportahan? Kaya sabi niya noon, kailangan daw talagang magtiis. Siya na mismo ang magtitiis para mapa aral niya ako hanggang college.

Alam mo, doon yung pinaka una at pinaka matagal na magkasama kami sa isang bahay. Doon ko naramdaman na, ah, ganito pala yung kapag may magulang ka, na aasikasuhin ka. Yung damit mo, yung paggising mo. Dun ka talaga naramdaman na may nanay ako. Tuwang tuwa ako!

Kasi di ko naranasan noong maliit pa lang ako, Ang paborito kong bonding namin ng mama ko sa Jeddah ay tuwing nagpupunta kami sa Jollibee. Kasi diba nakikita natin sa Jollibee, yung may bata doon, na sinusubuan ng magulang. Yun! Magtuturo ako sa Jollibee. Kunwari na ako, bata pa, baby ako. Tapos, sinasabi ko, Ma, paano ako? Kunwari baby pa ako. O, sige paano mo ba ako papakainin, Ma? Paano mo ako susubuan? Pinapakita naman niya sa akin.

Dalawang taon na ako dito sa Maynila mula noong pagbalik ko. Sa katapusan, aalis na uli ako pabalik ng Jeddah. Dala siguro uli 'tong backpack. Medical Lift Operator naman. Hindi na driver. Mas-safe yun.

26 For full text of monologue, see Geraldine Pratt and Caleb Johnston, in collaboration with the Philippine Women Centre of British Columbia, "Nanay (Mother): A Testimonial Play," in *Once More, With Feeling: Five Affecting Plays*, ed. Erin Hurley (Toronto: University of Toronto Press, 2014), 59–61.

27 **CHORUS**:

Tayo ay taga-Bagong Barrio.
Alam niyo ba kung ano ang hubog ng ating nakaraan?
Nung 1970s, ipinaglaban natin ang ating mga karapatan sa lupa, bahay at tubig.
Pitong daang katao ang nagtipon tipon noon upang magaklas.
May trabaho tayo noon at hindi gutom ang ating mga pamilya.
Hanggang sa pagdating ng salot na kontraktwalisasyon at paglisan ng mga pabrika patungong export processing zones.
Puro na lang tayo trabaho. Trabaho dito, doon. Kahit ano, kahit saan. Tayo ay nagtrabaho.
Sa dilim, naduduling tayo sa kadidikit ng bag na papel para sa National Bookstore.
Sumasakit ang mga daliri natin sa pagbabasahan!
Ang likod natin sa pagbabalat ng bawag para lang sa 70 pesos kada bag.
Gamit ang mapanglikhang isip, nakakagawa tayo ng trabaho kahit mula sa ere.
Loob natin ay hindi panatag,
walang anu mang kasiguruhan sa trabaho ang meron tayo,
Kaya ang iba sa atin ay lumisan patungong ibang bansa para lang mabuhay
Kaya tayo nag magsuri at mag organisa
Kapag tayo ay magsamasama, tayo ay ligtas sa kapahamakan
Mayroong kasiguruhan sa piling ng masa
Hindi sa lupa ng dayuhan ngunit dito sa ating pinakamamahal na bayan

28 Remembering the union activism and widespread civil unrest before and during the turbulent years of martial law (1972–1981) under President Marcos, Marilou, a long-term community organiser with PAMBU, an organisation of the working poor, recalled:

> The Care Jeans [factory] is where I spent nights on strike in the picket area. All of that I already wrote down [but lost] because, when this area became a hot spot, we had to bury those papers. We organized many workers [...]. The nuns. The priests. We were there with [the strikers] because they were already being harassed. It was real [...]. When the police were there, and when night comes, or early morning, they would destroy the picket line. They would have things to hit you with. It hurt. We would be sleeping and resting [...]. Then, suddenly, goons would approach. But, when the workers were organized and educated, they would fight. They would stand up. They would rebuild the picket area. They would continue. Continuous. They fought the struggle.

29 The Herrera Law amended the Philippine Labor Code of 1974 to create the legal grounds for contractual work arrangements, effectively breaking unionised labour and workers' association. Herrera introduced the concept of Assumption of Jurisdiction, "special" measures that curtail the workers' right to protest, allowing for the swift deployment of police and/or military to suppress and "resolve" industrial disputes. Former Sen. Ernesto Herrera has subsequently argued that his law has been misunderstood and that an earlier martial-law era Article 106, Presidential Decree 422, was already used by the Department of Labor and Employment to justify job contracting.

30 Rene E. Ofreneo, "Precarious Philippines: Expanding Informal Sector, 'Flexibilizing' Labor Market," *American Behavioral Scientist* 57, no. 4 (2013): 420–443.

31 Shannon Jackson, *Social Works: Performing Art, Supporting Publics* (New York and London: Routledge, 2011), 180.

32 Jasbir Puar, Lauren Berlant, Judith Butler, Bojana Cvejic, Isabell Lorey, and Ana Vujanovic, "Precarity Talk: A Virtual Roundtable with Lauren Berlant, Judith Butler, Bojana Cvejic, Isabell Lorey, Jasbir Puar, and Ana Vujanovic," *The Drama Review* 56, no. 4 (2012): 168.

33 Judith Butler, *Notes Towards a Performative Theory of Assembly* (Cambridge, MA: Harvard University Press, 2015), 11.

34 Ibid., 16.

35 Shannon Jackson, *Social Works: Performing Art, Supporting Publics* (New York and London: Routledge, 2011).

36 As Ola Johansson notes, this is often the case in assessments of community theatre. See Ola Johansson, "The Limits of Community-Based Theatre: Performance and HIV Prevention in Tanzania," *The Drama Review* 54, no. 1 (2010): 59–75.

37 Judith Butler, *Notes Towards a Performative Theory of Assembly* (Cambridge, MA: Harvard University Press, 2015), 20.

38 Jacqueline Lo and Helen Gilbert, "Toward a Topography of Cross-Cultural Theatre Praxis," *The Drama Review* 46, no. 3 (2002): 47.

39 See Diana Taylor, *The Archive and the Repertoire: Performing Cultural Memory in the Americas* (Durham and London: Duke University Press, 2003); Diana Taylor, "Save As," *On the Subject of Archives* 9, no. 1 & 2 (2012): n.p.; Diana Taylor, "Archiving the 'Thing': Teatro da Vertigem's Bom Retiro 958 Metros," *The Drama Review* 59, no. 2 (2015): 58–73.

40 Geraldine Pratt, *Families Apart: Migrant Mothers and the Conflicts of Labor and Love* (Minneapolis: University of Minnesota Press, 2012).

41 For details, see Geraldine Pratt, Caleb Johnston and Vanessa Banta, "Lifetimes of Disposability and Surplus Entrepreneurs in Bagong Barrio, Manila," *Antipode* 49, no. 1 (2017): 169–192.

42 That said, the first scene is an amalgam of two families' testimony.

43 Neferti X.M. Tadiar, *Things Fall Away: Historical Experience and the Making of Globalization* (Durham and London: Duke University Press, 2009), 225.
44 Marco Garrido, "The Ideology of the Dual City: The Modernist Ethic in the Corporate Development of Makati City, Metro Manila," *International Journal of Urban and Regional Research* 37, no. 1 (2013): 165–185.
45 Neferti X.M. Tadiar, *Things Fall Away: Historical Experience and the Making of Globalization* (Durham and London: Duke University Press, 2009), 144.
46 Neferti X.M. Tadiar, "Life-Times of Disposability in Global Neoliberalism," *Social Text 115* 31, no. 2 (2013): 43.
47 Gerald Vizenor, *Manifest Manners: Narratives on Postindian Survivance* (Lincoln: Nebraska, 1999).
48 Global Commission on International Migration, "Global Migration for an Interconnected World: New Directions for Action," Report, Switzerland, 2005.
49 Elizabeth Povinelli, *Economies of Abandonment* (Durham: Duke University Press, 2011).
50 Lauren Berlant, *Cruel Optimism* (Durham and London: Duke University Press, 2011).
51 See Robyn Magalit Rodriguez, *Migrants for Export: How the Philippine State Brokers Labor to the World* (Minneapolis and London: University of Minnesota Press, 2010).
52 Dylan Rodriguez, *Suspended Apocalypse: White Supremacy, Genocide, and the Filipino Condition* (Minneapolis: University of Minnesota Press, 2010), 186.
53 See Bobby Benedicto, *Under Bright Lights: Gay Manila and the Global Scene* (Minnesota: University of Minnesota Press, 2014); Neferti X.M. Tadiar, *Things Fall Away: Historical Experience and the Making of Globalization* (Durham and London: Duke University Press, 2009).
54 Dylan Rodriguez, *Suspended Apocalypse: White Supremacy, Genocide, and the Filipino Condition* (Minneapolis: University of Minnesota Press, 2010), 199.
55 Eve Tuck and K. Wayne Yang, "Decolonization Is Not a Metaphor," *Decolonization: Indigeneity, Education & Society* 1, no. 1 (2012): 1–40.
56 Grant Kester, *The One and the Many: Contemporary Collaborative Art in a Global Context* (Durham and London: Duke University Press, 2011), 115.
57 Denise Ferreira da Silva, *Toward a Global Idea of Race* (Minneapolis and London: University of Minnesota Press, 2007).
58 Dylan Rodriguez, *Suspended Apocalypse: White Supremacy, Genocide, and the Filipino Condition* (Minneapolis: University of Minnesota Press, 2010), 199.

4 Migration, settler colonialism and other futures

By 2014, one of the actors in three previous productions of *Nanay*, Hazel Venzon, had moved to a small city in the Canadian North, where she worked as a producer and programming associate at the Yukon Arts Centre. At her invitation and instigation, in January 2015, *Nanay* was presented as a staged reading as part of the Whitehorse Pivot Theatre Festival. For this performance, we worked with Hazel, local actors and community members to recalibrate the script to include a monologue of a migrant coming to Canada through another temporary migration programme, as a low paid service worker. Before making her way to Yukon, this woman had worked diligently as a housekeeper at a large hotel in the province of Alberta, and she had been told by her employer that they would nominate her for permanent resident status at the end of her two-year contract as a low-skilled temporary foreign worker. Rather than honouring this commitment, at the end of the two years, the employer put the names of all of their 50 temporary foreign worker employees "in a bucket" to choose five to nominate to the provincial nominee programme for permanent resident status. "I was crying," she said. "We gamble our life here," only to have one's fate left to another whim of chance. The employer did not choose this woman's name, and what followed was a period of undocumented status and living on the streets without work before finding her way to Whitehorse. Temporary migration through low-skilled service work has been a more common experience than migrating through the LCP for Filipinos in the Yukon and the new verbatim monologue spoke to this local experience.

The reaction to this script reading in Whitehorse was mixed and that of Filipino audience members differed from reactions in other places. Though attempts had been made to get the community out, Filipino presence was sparse that day and response tended to counter the themes and tones of *Nanay* by emphasising the successes within the Filipino community: "I look around," said one Filipino audience member: "They're [Filipinos] actually doing good. They're driving their own car. They're renting and eventually buying their own house. They work two jobs. They work three jobs. They're here. They're willing to work and contribute to the society and community. They're very adaptable. They adapt. And I think that's one of the greatest assets [of Filipinos], because they can adapt." Government officials present were even more expansive in the talkback after the reading. One stood to offer this commentary:

> I can't speak for the federal government, but I really liked the play. And I found it really interesting in terms of the grey areas because integration is, you know, a big adventure in life. And with adventure comes some pretty major challenges and I would suggest that just about everybody in this room, with the exception of I know of at least one person, has some kind of tie to immigration, in terms of whether their parents, their grandparents, their great great grandparents. And those stories, when you reach into your family history, are built with, you know, adventure. And often very difficult challenges. And I know with my grandparents – every one of them was an immigrant and every one had a really challenging story.

This speaker was followed by the individual who the government official had identified as the only non-immigrant in the room. "My name is Sharon," she said.

> I'm a Yukon First Nation, and I want to share my perspective, not on behalf of the eight thousand First Nations who live here, but from what I've seen with the community moving here. [...] I realize that as First Nations of the land, that we haven't continued our practice of welcoming people into our territory. Because of all the hardship since the Gold Rush and the [Alaska] Highway coming in, I think we're not as welcoming as we could be. But I would like to find a way to rectify that, to do our traditional welcoming. Thank you for your stories. One of the things that I heard was that family disruption. I was like, wow that sounds like what happened with residential schools. And I'm sure there has to be a way to dialogue that, to share with each other, to have the compassion for each other, and to make each other human. So my First Nations community, what I'm seeing [...] what I'm disturbed about is racism against immigrants, in particular Filipinos. And I try to fight against that but it's something that's very ... something that I want to stop in my community. And I think if people could maybe sit in a circle and tell stories, as is our way, that would be healing.

What emerged from these audience reactions was a new collaboration, in the first instance, between Sharon Shorty, the Tlingit elder and well-known storyteller and performer who stood up that day, and Hazel Venzon, the Filipina actor and producer who organised the staged reading in Whitehorse. Through a short artist residency in Juneau Alaska in December 2015, Sharon and Hazel conceived a new piece called *Tlingipino Bingo*, which was collectively built and performed at Whitehorse Nuit Blanche in June 2016. It is this spin-off from our original play that we examine here, to explore ways of responding critically and analytically to Sharon's invitation to share stories and build relations on terms other than those on offer that day, that is, to build relations that move away from a model of minority immigrant success within narratives of state multiculturalism, towards new ontologies of belonging and social relatedness, including familial relations imagined and lived beyond the private family. In seeing the basis for conversation across First Nations' experiences of state-enforced family separation through residential schools and Filipina/o labour migrants' experiences of family separation, Sharon Shorty joins and brought us to ongoing debates about the relationships between migration and Indigenous studies, race and Indigeniety, racial global capitalism and settler colonialism, and the possibilities of relationship between Indigenous peoples and those forced into the Americas by the violence of European and Anglo-American colonialisms and imperialism around the world.[1]

As white settlers, we sat awkwardly within this collaboration – differently awkward than was the case in previous productions (Figure 4.1). *Nanay* originated as a staged dialogue between Filipinos and white employers and the Canadian state. With the PETA production, dialogue moved to cross-class and transnational conversations. In Bagong Barrio, the collaboration with Migrante allowed us collectively to situate Canadian migration policy within a broader geopolitics of uneven development, and focused attention on decolonial disruptions in relation to U.S. imperialism. *Tlingipino Bingo* staged storytelling across Indigenous people and Filipinos in Whitehorse, in the context of a settler colonial society. As backdrop to this conversation between Tlingit and Filipino migrants, our positionality and role within the process was both ever present and less specified.[2] The creation of *Tlingipino Bingo* was a productively discomforting process sitting within an Indigenous-arrivant-settler triad. This is no less the case for this critical reflection. We document the public event of *Tlingipino Bingo* (and draw on additional research materials from Whitehorse) to interrogate how deeply settler colonialism burrows into everyday life, including – perhaps – practices of racialised immigrants, the ways that a model minority discourse functions within state multiculturalism, and to imagine other futures beyond settler colonialism, which could possibly include us as allies. Not only does it challenge and support Filipino migrants to recognise belonging beyond inclusion within the conventional norms of citizenship, embedded within the specifics and logics of settler colonialism and the racial dynamics and concerns of a small city in Northern Canada, *Tlingipino Bingo* possibly also helps us all to think strategically about possible critical responses to contemporary claims of dispossession by white citizens in Canada and elsewhere, as well as their destructive nostalgia for a lost time of national whiteness.

Setting the stage: Filipino labour migration and settler colonialism

Sharon Shorty's observation about antagonism between local First Nations and Filipinos in Whitehorse was mirrored in conversations with some

Figure 4.1 Conversation Collective and flyer.

(not all) Filipino residents. Some spoke from their own experience: "I was walking along Third Avenue, and then these two First Nations were coming towards me and I was putting coins in the meter, and they yelled at me: 'Go back to where you came from.'" More common, stories of negative encounters circulated as hearsay, along with rumours of a Facebook page in which hostile comments could be viewed. Conversely, some Filipino community members voiced their own stereotypes and resentment about First Nations' peoples. "They call them *pana. P-a-n-a. Pana.* So that nobody knows that we're talking about them. And there's some resentment that they get more, they have more. You know, when they're in school they get this and they get that." Several Filipino residents responded to the accusation that they have stolen jobs by calling up their capacity as workers and their obligation to family: "I think there's jealousy because you see so many Filipinos driving brand new cars, owning a house, so progressing in their lives because, first and foremost, they came here for economic reasons. [Enjoying life is] not what we came here for. It's not just to get by but to provide for ourselves and then help other family members. And that's the strength of the Filipinos." From another Filipino community member we heard, "Some of the First Nations would say: 'You're taking away jobs from us.' But what I've heard from employers is: 'How could we employ them, if they are not educated for their jobs? Like we will train them for a few weeks, and then after a few months they will just say, 'Goodbye. I can't work for you anymore because there are personal reasons.'" This community member noted that Filipinos are willing to work diligently at low-level service jobs even after they have fulfilled the terms of the immigration programme that brought them to Canada: "Especially under the Yukon Nominee Program, although they are required to stay for two years [with the employer who sponsors them], many of them stay working with them longer than their contracts. Some of them came in 2007 [to work as clerks for Canadian Tire] and they are still there." One Filipino resident reasoned that one implication of this capacity for hard work and reliability as workers is that there is often little time for Filipino immigrants to learn about the history of settler colonialism in Canada:

> Not a lot of Filipinos really know the story of First Nations [...] A lot of Filipinos don't have time to read or to go to college. Because they're here to work. Time is money. Per hour. 'If I could fit 3 jobs into 24 hours, that's great.' Two jobs maybe. So they're here to work. They're here to make money. It's not monthly [...] In the Philippines you're paid by the month. By the week, or twice a month. Here it's per hour. Per hour for them is a new notion. It's like, 'Oh my god, I can make money. Like a lot of money. Per hour.'

The same community member suggested further that even when histories of colonialism are known by Filipinos, there may be reluctance to assume responsibility for colonial legacies and traumas in the Canadian North because they predate their arrival.

These observations among Filipinos are embedded in stories about two phases of Filipino immigration to Yukon and within a context of a significant Indigenous presence (unlike many metropolitan areas that have been significant sites of Filipino settlement in Canada, such as Vancouver and Toronto). More than half (54%) of Indigenous people in Yukon live in Whitehorse and represent 18% of the city's total population.[3] Considering Filipino migration to Yukon, the first phase is narrated around seven successful nannies, whom we heard about again and again during our time in Whitehorse.[4] This story was so common that when we mentioned it to a non-Filipino government official, he said, "Oh, so you've heard the story." The narrative is about seven single Filipina women, friends who attended the same church while working as domestic helpers in Singapore in the 1980s. One found employment in Yukon as a live-in caregiver in 1984, and by the early 1990s the other six had joined her in Yukon, also as live-in domestic workers. These seven women have been successful in many ways, including purchasing houses and, in some cases, starting businesses. One of the successful nannies told the story of property ownership in this way:

> 1991, 1992: I called [three other nannies] one day and said, 'Let's put our money together and buy a house.' Because we were just nannies, right?, we were kind of jokingly saying, 'Are you dreaming or hallucinating?' And I said, 'There's no harm in trying. We can just check it out.' I remember we got dressed up, because we really wanted to represent ourselves. We go into the CIBC [Canadian Imperial Bank of Commerce]. It was really easy. We never thought [we would get a mortgage]. At that time we were being paid 6.50 an hour. So we put our salary [together]. And then we were laughing because when we came out – after 15–20 minutes – we were pre-approved for [a] $160,000 [loan]. And we never thought [that was possible]. We thought it would be so hard! But then, anyway, we started being picky because we realized how easily we can come up with the money. So we went to the other bank as well, because at the CIBC there was a lady there that kind of treated us differently. And then somebody told us, 'You guys, you're really not asking their favour. You're giving them business, eh?' So it's something that we have to be proud about. But being a nanny, it's so hard to put that in your mind. We were the first Filipino nannies buying a house. So that kind of opened an inspiration to others. Two years later on, two partners went out and bought their own place too, and it started just stirring up inspiration for others to buy a house. [...] Same thing with the car.

They became known as the "Teslin girls" for the street on which their house was located and their home would become a gathering place within the community.[5]

One of the seven nannies' most influential successes has been to sponsor large numbers of family members since 2007 through the Yukon Nominee Program (YNP). This is an immigration programme that allows employers, who can establish to the government that they cannot find Canadian citizens or permanent residents to fill their jobs, to nominate non-Canadians to do this work. Workers come under a temporary work permit and have the opportunity to apply for permanent resident status within the YNP, at which point family dependents also can be sponsored. A government official associated with the programme confirmed that until relatively recently the programme was run by two people "kind of off the side of their desks." He acknowledged that several of the successful nannies had seized the opportunity of an initially under-regulated immigration programme by approaching Whitehorse employers to say "Looks like you need some people working in your business," presenting resumes of family members and doing the necessary paperwork to enable these employers to nominate their relatives through the YNP.[6] Successful nannies' creative manoeuvring within a formerly loosely regulated programme, which is explicitly geared to employer need and labour market requirements, allowed them to sponsor many family members relatively quickly and speaks eloquently to Nina Glick Schiller's and Ayse Caglar's insistence on the agency of migrants in processes of urban growth, and in particular their role as labour recruiters, especially in smaller (what they term "low-scale") cities.[7] Between 2007 and 2016, 410 workers from the Philippines and 382 of their dependents came to Yukon through the YNP, almost all as low-skilled (or what the government calls "critical impact") workers (food counter attendants, light duty cleaners, cashiers, sales clerks and store shelf stockers are among the most common occupations included within this designated classification). Repeating a familiar pattern of racialised labour migration, Filipinos are noticeably over-represented among critical impact (as compared to skilled) workers: half of all critical impact workers are from the Philippines, and only 12% of Filipinos come as skilled workers as compared to 68% of those coming from places other than the Philippines.

The effects of the YNP are highly visible in the Whitehorse landscape. In the words of a local Filipino resident, "You know the hotels, it's all Filipinos now. The restaurants, the fast food, the Canadian Tire, Home Hardware, Superstore. It's all Filipinos. McDonald's. It's all Filipinos. Tim Hortons…." One government official estimated that Filipinos now comprise between 5% and 10% of the Whitehorse population. There is now a competitive Filipino basketball league in Whitehorse. There are Filipino restaurants (e.g., The Talk of the Town restaurant) and remittance-sending agencies. The Superstore stocks an array of Filipino foods, everything you need to make *halo halo*. The *Philippine Asian News Today* is widely available, and there's a small kiosk in a local mall where one can buy and send *balikbayan* boxes or directly purchase gifts – flowers or a birthday meal at Jollibee's – for one's family or children in the Philippines.

Less visible is the creation of what is regularly referred to as clans within the Filipino community. One of the successful nannies estimates that she has helped to bring more than 30 family members, who have then sponsored their dependents. Altogether she estimates that she has helped 75 relatives and people from her village through the YNP. Speaking of a number of the successful nannies we were told, "They just brought the barangay [village] here." An article in a national magazine, *The Walrus*, states that one of these women almost "single-handedly transplanted a community from the Philippines to Whitehorse."[8] Several Filipino community members speculated that the successful nannies' agile use of the YNP has defined social and economic relations within the Filipino community in Whitehorse, in both positive and potentially troubling ways. Without necessarily leading to this outcome, it has created structural conditions for possible exploitation within Filipino families, as newcomers are obliged to repay their debt of gratitude to the family member who arranged their sponsorship.[9] Some Filipino community members also noted that it potentially creates a pool of quiescent workers in low-level service jobs under conditions where it is difficult to challenge an employer, because this would bring shame to the family member who recommended them for the job.

How Filipino migration to Yukon can be thought in relation to settler colonialism is an open, contested question and we acknowledge that we enter this debate as white settlers. Within debates about the relations between settler colonialism and racialised labour migration, one point is clear: these experiences cannot be collapsed into each other. Racialised immigrants' experiences of labour exploitation and Indigenous peoples' dispossession from land and resources are distinct processes within global capitalism. The logics of racial exclusion are different from the logic of elimination, and the possibilities for solidarity lie in what Tuck and Wang have termed an "ethic of incommensurability."[10] Jodi Byrd urges that we read the "cacophonies" of colonialism as they are rather than attempting to hierarchise, equate or place them in causal order.[11] The processes are nonetheless conjoined in specific histories of violence in specific places: genocide and dispossession cleared the land for the use of indentured labour; and indentured labour was in some places introduced as a kind of solution to political pressure to abolish slavery.[12] Jodi Byrd argues that ideas of Indian and Indianness have functioned as the "transit" of U.S. empire, the "ontological ground through which U.S. settler colonialism enacts itself as settler imperialism,"[13] a claim that demands close attention in the Philippines, where the U.S. colonial experience has been so formative. For example, modes of warfare and actual high-ranking U.S. army officers from the wars fought with Plains Nations in North America were directly transported to what Hedman and Sidel term "the United States' first armed adventure in Asia"[14]: the Philippine-American War 1899–1902. Neferti Tadiar documents how the incorporation of the Philippines into U.S. domestic space as a colonial acquisition "bore the legal memory" of earlier landmark legal cases framing citizenship for African and Native Americans in the United States.[15] In not extending the Fourteenth Amendment of the U.S. Constitution to the Philippines as an unincorporated territory in 1901, "we could say Filipino is racialized as *not* black, and *like* Indian" at one defining moment in its national formation. As in the case of Indigenous peoples and slaves, the withholding of U.S. citizenship to Filipino colonials, she argues, "set precedents for deeming what is outside the national rule of law as also outside the bounds of civil and human status, which continues to be reflected in the attitudes today towards 'illegals' (as forms of non-personhood)."[16]

Alongside these interwoven histories that create the ground for solidarities, immigration can and often does reinforce the colonial and multicultural state. As Patrick Wolfe has argued, there is nothing preventing even colonised natives from one region becoming settlers in another if their actions support the dispossession of Indigenous peoples in the new locale.[17] Dean Saranillio also notes, "Power does not simply target historically oppressed communities but also operates through their practices, ambitions, narratives and silences."[18] Is it significant that the seven successful nannies are regularly referred to as pioneers within the Whitehorse community? Do Filipino homeownership and incorporation within circuits of finance capitalism signify processes beyond immigrant success? Have the seven successful nannies delivered almost a thousand well-educated, quiescent workers to Whitehorse service industry employers, even as they exercised their ingenuity to reunite their extended families and barangay relations in Yukon? Is the following a harmless slip? "We were joking that there's only between 29,000 to 32,000 in a land area three times bigger than the Philippines. We are 98 million at present in the Philippines. [...] I said, 'There's still space that we can fill in.'" The seven successful nannies' and other Filipino immigrant stories in Whitehorse create uncomfortable, immensely productive moments in which the figure of Filipino-as-victim of empire crashes up against Filipino-as-settler in profoundly destabilising ways.[19]

As non-Filipino, non-Indigenous scholars, we are drawn to comment on these narratives because they do considerable ideological work beyond the Filipino community. We heard them from non-Filipino government officials and they are widely circulated in the local and national media.[20] Hard-working Filipinos, working 2–3 jobs to amass all that is possible in 24 hours on an hourly wage, persevering in their jobs at Canadian Tire long after their temporary work permit requires it – this fits within a long history of configuring Asian workers within a "natural" appetite for hard work. The hard work, sacrifice and success of Filipinos in Whitehorse are not at issue. However, this discourse does significant ideological work within the wider Canadian society. It marks Filipino racial difference and renders Filipinos as hyper economic embodiments of laissez-faire capitalism. Iyko Day traces in the Canadian context the extent to which this portrayal of Asian labour has both marked Asian labour as "model minorities" and historically been a source of criticism from white workers, who feel that Asian workers' appetite for self-exploitation has undermined white workers' wages and working conditions and capacity for human and humane conditions of social reproduction.[21] In other words, historically this appetite for hard work has been used to render Asian workers as somewhat less than fully human. Stories of hard-working immigrants' success simultaneously sustain a powerful liberal narrative of the racial neutrality of the state and capitalist economy (i.e., we all can be successful regardless of race), as well as a narrative of equality and progress towards a post-racial society.[22] Both the Canadian and Philippine states have considerable investment in this discourse: it ties worthy citizenship to the "good worker" and remittances from these conscientious workers sustain the Philippines' economy.

Played out directly in relation to Indigenous Canadians, the narrative of hard-working Filipinos does further ideological work. By rendering Indigenous peoples as indigent and incapable of disciplined labour in invidious comparison, it can breathe new life into an argument that laid the philosophical foundations for British claims to property in North America and justified dispossession of Indigenous peoples from their lands: this is the Lockean argument that appropriation through labour underpins rights to ownership. The complexity of Filipinos' positioning within settler colonialism brings all of us to the reality that settler colonialism is a violent *ongoing* process, often unwittingly enacted by sympathetic well-meaning people with complicated lineages and relationships to it.[23]

Tlingipino Bingo entered into this fraught politics and worked with these complexities and complicities in ways that we turn to discuss. Through our presentation of the performance, we hope to lay out some strategies for a prefigurative politics beyond liberal forms of inclusion and exclusion, including perhaps, alternative narratives of Filipinos beyond self-exploiting model minorities.

Joking across colonial traumas

Tlingipino Bingo was an interactive improvised bingo game performed at the Elks Lodge 306 in Whitehorse as part of an all-night arts festival. The bingo game provided the structure within which three performers improvised. It grew out of Sharon and Hazel's discovery during the artist residency of a shared enjoyment of bingo within First Nations and Filipino communities, typically played, however, in different places in Whitehorse: at the Elks Lodge (and as TV bingo during the winter months) for the former and at private parties for the latter. Their objective was to work with their communities' mutual enjoyment of this game of chance to get them into the same space to play together and share stories. We arranged for one "plant" in the audience, but much of the performance was unplanned, itself left to chance and the improvisational instincts of the performers. We put together a series of prizes – bags of rice, a basket of ingredients to make moose adobo, a *balikbayan* box, Chilkat woven earrings, a bucket of Kentucky Fried Chicken, Aunty Wilma's canned salmon, among other prizes that we hoped would prompt the sharing of stories.

The Elks Lodge runs a bingo every day of the week, twice on Saturday and Sunday. Each game takes about two hours, but many bingo players come hours in advance to claim their lucky chair. Roughly 140–160 people come out for each game, the same 75 to every game, and some players travel 100 miles to play. In terms of the bingo players, it is an Indigenous space: an Elk who has been running the bingo for 40 years estimated that 60%–70% of the players are members of First Nations. We scheduled *Tlingipino Bingo* right after the Saturday night game to keep these players in the audience and worked with the Canadian Filipino Association of Yukon (CFAY) to bring the Filipino community out.

The Master of Ceremonies was Sharon Shorty, performing in character as Gramma Susie. Gramma Susie "came to life" in 1996 and was intended as a one-time performance with her friend, Jackie Bear, another Whitehorse Indigenous artist. They brought together their shared history of being raised by their grandmothers and improvised their characters, Susie and Sarah – "two old ladies always laughing about something" – as a way of respecting and honouring their elders. Gramma Susie has lived on and Sharon is now a well-known and well-respected award-winning performer and storyteller, who performed at the Truth and Reconciliation National Event in Winnipeg in 2010 and was part of the Yukon First Nations contingent to the 2010 Winter Olympic Games, among other notable performances. There is a display of her iconic outfit at the Museum of History in Ottawa, with each item carefully catalogued in its digital archive. Gramma Susie was, in other words, a familiar character to at least a portion of the audience at *Tlingipino Bingo*. She is "feisty, opinionated and funny," an elder who delights in cutting "big shots" down to size.[24] Of the territorial premiers who she has forced to learn the Indian dance over the last 20 years, she says, "Some of them are really scared of Gramma Susie because she's very politically astute. Knows what's going on. Knows their weaknesses." She uses her sharp wit to confront settler colonialism with humour, by telling stories about meeting the Queen, Prince Charles and Colonel Sanders, the latter a source of sustained romantic attraction (Figure 4.2).

Gramma Susie began the performance/game by taking note of her surroundings, facing off with a display of Elks at the end of the hall: one massive stuffed animal head amidst framed photographic images of past presidents of the Benevolent and Protective Order of Elks of Whitehorse, all white and

Figure 4.2 Gramma Susie and Ardie.

all but one male, not surprising given that the first women were admitted only in the 1990s and the all-white clause of the Order of Elks' constitution was revoked only in 1976, with the proviso that it could be reinstated if the law allowed. "For goodness sakes," Gramma Susie said, "I like these guys. They look like Grand Poobahs from – what do you call it? The Flintstones!" She welcomed the crowd by calling out first Filipinos, then First Nations, then artists, then Chinese and then Tlingit to identify themselves. As she was equivocating from her role as bingo caller, she detected the Filipino performer, Ardie, in the crowd and called out "Oh, look! There's Ardie from Tim Hortons" and brought him on stage. To the audience, "Listen, this is Ardie. He's my favourite server at Tim Hortons." To Ardie, "I didn't know they do delivery now! Wow. Just for Susie. Listen, Ardie, I'm pretty tired out and you're Filipino. You can handle one more job, can't you? Can you call bingo numbers for me? I'll hand out your timbits [bite-sized donuts]. I'll do your job, you'll do mine. Fair trade. Listen, I want to ask. Welcome to the Yukon, our home. This is our special cultural gathering. It's called a bingo hall, and that's where we hang out. Why did you come to Yukon?" And so it went, Ardie called the numbers and Gramma Susie handed out timbits and cruised the crowd for stories.

As chance would have it, the same white audience member won the first two prizes. After the first win, Gramma Susie told him: "Don't win anymore tonight." When he won again, she said, "I told you not to win again" and threatened to beat him up. Ardie intervened: "Gramma Susie, I think we have a special guest in the house tonight to share a story." Gramma Susie quipped: "Well, she almost won in the bingo but [the two-time winner] stole it. Take away his cards." "Now, listen," she said, "we just handed out some white rice [for a prize]. I was meeting my new friend… Marivic could you stand up?" She explains that she and Marivic (the woman identified by Ardie and the only "plant" in the audience) had been talking about what they liked. She likes the white man Colonel Sanders and she asked Marivic to tell the audience what she liked in the Philippines. Marivic responded with a story she had shared with Sharon in advance:

> When I was a little girl, I'm dreaming to be a white. I was bullied during grade school. They said I was so dark, so I was dreaming to be white. When I grew up and got a job, and I had a friend, a dermatologist, she said: 'I can make you white.' There's a pill I took, and after 7 days, I was white. But I didn't like it because I looked Japanese [not myself] and I decided I like me being brown.

"Well, look at that," said Gramma Susie, placing her arm alongside that of Marivic. "We're just about the same colour. Look at that. Beautiful, right? Thank you, Marivic, for sharing your story. Hopefully this time you get a good bingo." Marivic's story about her desire for and efforts to achieve whiteness is an intimate story about the enduring violence of U.S. colonialism

Figure 4.3 Sharing stories.

in the Philippines. The combination of armpit whitening and glutathinone tablets prescribed by her trusted dermatologist friend, widely used to alter skin pigmentation, is carcinogenic with additional side effects, including insomnia, nausea, vomiting, and damage to the liver (Figure 4.3).

Trading stories about desire for whiteness, whether it be embodied by Colonel Sanders or imbibed as potentially life-threatening prescription drugs, joking about Filipinos' labour exploitation in Whitehorse, these are small barbed critiques shared across distinctive colonial histories. When Gramma Susie first met Marivic, she shared other stories, including the memory among her people of when they first met white people, the pity they felt for the poor health associated with their pallor, and their efforts to restore white people to health by feeding them meat. Stories open to other stories and our collective hope was that the bingo performance would prompt audience members to reflect on and speak across different but connected traumas, and to think more fully about deeply troubling histories and experiences within two distinctive colonialisms. It is difficult to gauge to extent of reflection, but some white audience members were visibly shocked by some of the stories.

There is no depth of analysis attending this proliferation of stories – it's just joking around, with gentle teasing and prodding by Gramma Susie. There is a rich tradition of this kind of joking around in First Nations' literature and drama in North America, and considerable reflection on the political possibilities of reframing colonial encounters and racial stereotypes through humour. Speaking about the character of Gramma Susie in the film *Redskins, Tricksters and Puppy Stew*, Sharon describes laughter as the "greatest joy": "I often think it's a miracle to make somebody laugh. Because you have no idea what will spur that laughter on. And I've thought a lot about what humour is, and what we as native people think humour is. And a lot of times it depends on the chaos. If you look at our history, we've been through a lot of chaos as people. And sometimes chaos makes us laugh."[25] Humour nurtures resilience and creativity across colonialisms and laughing together (in the face of and, in some cases, at the expense of white settlers in the audience) works away at the mutual racisms generated across immigrant and Indigenous populations, and positions the white settler in a productively awkward relationship to the jokes.

The possibilities of sharing at *Tlingipino Bingo* went beyond storytelling because of the embodied interactive nature of the performance, of literally playing together. The director tells one small story that gives a sense of the possibility:

> [The Bingo performance had] a different way of connecting people. [...] Like, there was this aboriginal woman who came in late, and there's no more bingo cards. And it's her space. [...] I said, 'Just wait.' I went to a Filipino to ask him if he could give his card to this woman and he said, 'Yeah, sure.' And I called the woman and I introduced them, and he immediately gave the card. He said 'I can share one with my wife.' Then it becomes a family thing. Giving is important. And allowing yourself to surrender your power, your chance to win. Like you give it to the original people, because of who has been occupying their land. Anyway, it's fun. It's fun to give. It's fun to share. And then, the woman won. The woman won! [The Filipino man who] gave his card. He found me. 'Dennis, our card, it was a winning card!' And we're laughing. You could be happy even if you lost your chance. To be a witness of losing but still ending up happy and joyful because anyway he had moments with his wife: sharing, dabbing, looking for numbers. That community-ness is present.

Tlingipino drag and rubbing history against the grain

There was a short break after what were called the Filipino and Tlingit rounds of bingo. When everyone was seated once again, MIA's song "Bingo" blared from the speakers and Miss Lituya entered from the back of the hall, lip-synching and dancing her way to the front on four-inch high heels. The atmosphere was electrified, many phones held high in the air to record the performance, and there was an excited buzz of whispered questioning about Miss Lituya's gender. When Miss Lituya had made her way to the front of the hall, Gramma Susie observed: "You're a tall lady. Who are you anyway?" Miss Lituya explained: "My name is Lituya Hart. I'm a woman of the Coho clan, the clan house that faces outwards. My father's clan is the Wooshkeetaanyádi [Shark clan]. [Sharon:] You're my people: Tlingit. But, you're Tlingipino too." Ardie intervened: "I hear you are half Filipino. Do you speak Tagalog?" Miss Lituya's response that, "No. My great grandfather immigrated to Southeast Alaska from Ilocos" elicited loud cheering from the Filipinos in the room. And so Gramma Susie did a second round of welcoming: "So, here's Gramma Susie welcoming properly, first time ever in the history here at Elk's Lodge, our first ever Tlingipino drag queen." Turning to Ardie, Miss Lituya extended the offer of kin relationship: "Hey cousin, when you're done helping Gramma, could you come and help me call [the bingo], because I'll be delivering the prizes" (Figures 4.4 and 4.5).

Miss Lituya is the creation of Ricky Tagaban, a Tlingit artist and storyteller. He has given Miss Lituya a place within his own Tlingit family origins in Lituya Bay. As the fates would have it, Lituya Bay was the site of the largest recorded tsunami in 1958. "To see so much of your family go back to the land after a natural disaster like that," says Ricky, "it reinforces our claim to that place. It's a very Coho place." Ricky is also the great grandson of a Filipino who migrated to Alaska to work as a commercial fisherman.

Miss Lituya is a transformative figure, we suggest in three ways. First, she destabilises conventions of gender, sexuality and race, including white conventions of queer.[26] Hunt and Holmes argue that since gender binarism was inherent to the project of colonial erasure of Indigenous peoples, decolonisation is "already active" in the lived experience of Indigenous queer gender and sexuality.[27] Beyond this, securely situated within her Tlingit clan, Miss Lituya does not present in the first instance as queer (or simply so) and her drag character is not easily mapped onto white traditions of drag.

Writing about the Filipino tradition of *bakla*, Martin Manalansan argues that it is a mistake to interpret this form of cross-dressing within white drag, which tends towards the more parodic, scandalous and comedic.[28] For Filipinos, he argues, manipulation of surface appearance through cross dressing is not a parody, nor is it done to suggest a consistent gendered self or to unveil something that has been hidden. Rather, the "literal application of

Figure 4.4 Transformation.

layers upon layers of signs of beauty enacts an ontological transformation"; it is "playing with the world."[29] Manalanson is drawing on Fenella Cannell's ethnography of lowland Christian Philippines, in which she locates *bakla* beauty contestants within ontologies of spirit mediumship and relationships to power. In contemporary Philippines, such mediumship has now typically morphed away from spirit mediumship, she argues, towards mediating colonial relations through the power of beauty and the embodiment of American notions of glamour. *Bakla* beauty contestants "become the temporary bodily 'lodging places for potency' … [recapturing] power, not literally through possession, but through a wrapping of the body in symbols of protective status, and a transformation of the persona by proximity to the power it imitates."[30] Miss Lituya brought an aspect of this into her performance when, with the guidance of the director, Dennis Gupa, she created as the outer layer of her costume a jean jacket with an *agimat* (protective amulet) sewn to the back. She explained to the audience: "This is a contemporary *agimat*. [Interrupted from enthusiastic comments and explanations from Filipinos in audience.] Yeah. So it's … [More explanation for audience.] Right. So the intention when you wear this is… you don't get harmed by others. So it's for protection. And it's believed to be bullet proof. So it is kind of a sacred garment. And this is Lituya's version of it with the Tlingit leopards on the back of it" [Laughter and clapping from audience].

Drag characters have been deployed by other First Nations contemporary artists, also in transformative ways. In Canada, Kent Monkman, a prominent artist of Cree and Irish ancestry, displaces popular visual imaginary of Indigenous men as doomed lamentable "Noble Savages" and reanimates Canadian history through the antics of his drag character, who he both performs and paints. He has painted Miss Chief Share Eagle Testickle, for example, into Robert Harris's famous group portrait of the Fathers of

Figure 4.5 Transformation.

Confederation: "She's trying to get a seat at the table," he explains, "or she could be a hired entertainer."[31]

Miss Lituya disrupts national history in another way: she is living proof that there is nothing new about Filipino migration to the region,[32] potentially opening history and the narrative of the Canadian nation in provocative and promising ways insofar as the primacy of white over Filipino settlement is somewhat disrupted. Miss Lituya not only embodies this history herself, but other histories of Filipino-Tlingit exchange gathered around Miss Lituya and the project. We were surprised when we visited Teslin, the location of Tlingit self-government, at the beginning of our process, and were told a story by an elder about her trip to the Philippines in 1976. She visited for a month, mostly in Baguio, where she met with the faith healer, Agpaoa. She had especially fond memories of a trip to Manila, dancing with a Filipino man named Romeo at the All American Bar. Our surprise itself tells the story of "the colonization of memory and of events that come to be known as 'History.'"[33] Similar stories kept returning in different ways. At *Tlingipino Bingo*, this story told by an elder won the grand prize:

> Well, this is the most fun I've ever had in my life in the bingo hall. In 1977, I had the opportunity to go to the Philippines, and I went to Baguio City. And when I went there, everybody wanted to come home with me. [Addressing Filipinos in the audience:] It took a long time for you to get here but I'd really like to welcome you and tell you a story about what happened when I was there. My auntie Angela – some of you may know her – she went shopping. There's two things that happened. First of all, we were in a bus. You know in Baguio what it's like, and Auntie was about 77 at that time. Everybody was in the bus, and the bus stops and the bus driver said, 'Anybody want to go and take a look at the city?' My auntie says, 'I do.' So, she gets off and everybody was shocked. 'Where is she going?' Anyway, she got off and went shopping, and she goes into this store and says 'I want to buy a dress.' They say, 'What size?' 'You know my niece Shirley? She was here 1977.' This was about 1979, and they say, 'Yes.' Which they didn't. 'I want to buy a dress to fit my daughter.' So she bought this dress and it fit her perfectly. [...] That was one story about Auntie Angela. And the other thing that she did was you know in 1979 when she came back, her daughter asked her, 'Mom how come you're dressed so crazy?' She had 4 pairs of pants on. She had t-shirts, shirts. I don't know what else she had under her clothes. I just said to her, 'How come you're dressed like this?' 'Well they told me I can't bring things across that border so I smuggled it across.' That was us First Nations in your country [the Philippines].

We began an interview with the first president of the CFAY by explaining that our theatre piece brought the Filipino and Tlingit communities together to play bingo.

> "Tlingit?!" she said. "Why are you…?" We started to explain and she interrupted "But did you know that I'm a Tlingit?" A single mother of three children living in Baguio, she met a visiting white Yukoner, and eventually married and came to live with him in Yukon. "When I arrived here," she told us, "my second boy opened a restaurant […]. So the name of the restaurant is William's Place because my father's name is Williams. And I was carrying that as my maiden name. So this lady comes to the restaurant. And then she said, "Who is Williams?" Because that's her surname [too]. I said "Me. My son is Williams." "Why are you Williams?" "Because," I said "my father is a Williams." She said, "Oh my god, where are you from originally?" I said from the Philippines. "Oh my god," she said. "You are the one we're looking for" [thereby introducing her to a history she did not know]. Isn't that a coincidence that I married a guy from here, that's living here? So what they do is they take me to the Indian Affairs and they look at the book, the big book, and it's written. Girl. Philippines. So, I became status Indian. And all my children are status because I'm not married to the father of the kids. So my status, I'm 50% Indian. My children carry it."

Her Tlingit father met her mother in Leyte in the Philippines while serving in the Canadian navy. On the one hand, this history points up the gendered dynamics of sex and race within settler colonialism: under the Indian Act, non-Indigenous women could acquire Indian status through marriage to an Indigenous man and transmit status to their children, while the same was not possible for a non-Indigenous man marrying an Indigenous woman. Indigenous women marrying non-Indigenous men lost their status, with profound effects on gender dynamics within their own community.[34] At the same time, as this woman's story retells this familiar narrative of patriarchal settler colonial legal ordering, there is something almost fantastical about it, told in a remote settlement more than an hour outside Whitehorse, in which the world is suddenly not quite as we expect. In a Benjaminian sense, it momentarily breaks the spell of ossified tradition "by cutting out what is 'rich and strange' … from what had been handed down [as History] in one solid piece."[35] This is a Filipino woman being found within a history of which she knew nothing, one that embeds her within the contemporary Tlingit community and confers rights to resources (e.g., a share of fish) and care (e.g., shovelling her snow) that have nothing to do with Canadian citizenship.

These stories are disruptive of colonial framings in a number of different ways. They surely disrupt static narratives of Indigeneity. Tlingit people in these stories are travelling the world in search of healing and as tourists or while enlisted in the Canadian navy.[36] It reclaims, in Audra Simpson's words, "fluidity of traditions and histories" that persists within and beyond classifications and "rigidities of colonialism."[37] They are in the world, in history, in time.[38] The stories gesture to rich histories, unknown to Filipinos and some Tlingit in Whitehorse, of already existing relationships between

Filipinos and Tlingit in the region, going back to 1791,[39] and in recent memory with the arrival of Filipino men in the 1920s and 1930s, who had migrated north – typically as contract labourers – to work in the seasonal Alaskan canning industry or as commercial fishermen. Many stayed, some (like Ricky's great grandfather) married into First Nations communities. Filipinos now represent the largest non-aboriginal ethnic group in Alaska (4.4% of the population according to the 2010 U.S. census).

Filipino and Tlingit histories do not blur into each other, but there is a complex set of intimacies that tends to be forgotten or written out of accounts; indeed, Renisa Mawani traces the efforts on the part of the Canadian state to restrict such intimate connections.[40] Silences about histories of international intimacies, like the history of Filipinos and Tlingit, stop us from knowing other ways of relating, and different pasts that are also resources for imagining different futures. Reclaiming histories can open up a new political imagination based on a "contextualized and historicized understanding of the relationships between the different peoples who have interacted and co-existed in places over time."[41] Knowing these histories releases into the space of colonial history, "other normative and theoretical thought enshrined in other existing life practices and their archives." "For it is only in this way," Dipesh Chakrabarty writes, "that we can create plural normative horizons specific to our existence and relevant to the examination of their lives and their possibilities."[42] It also creates a space for nurturing contemporary solidarities and intimacies among those who are torn apart by citizenship and by multicultural inclusion and its violences within a settler state (Figure 4.6).

And finally, Miss Lituya's beckoning and recognition of Ardie through the familial term, "cousin," is a solicitation, not from within the bourgeois private family, but through a different ontology of social relatedness,[43] one that Filipino audience members may have been able to hear in these terms. While installing the heteronormative family and norms of private domestic life was central to the U.S. government's colonial practice in the Philippines, Filipino-American scholar, Neferti Tadiar, has argued that less individualistic and bounded ontologies of self and social belonging continue to "cut across and confound" these outcomes.[44] She finds, for example, in Filipino migrant domestic workers a means of imagining forms of sociality that are illegible in typical renderings of them as racialised and exploited workers. As "conduits of other people's wills and aspirations, accommodating and conforming to the bodily, emotional, psychical, and metaphysical requirements of individuals and communities, to which they are attached as vital, component yet alienable parts, they act in a very practical but also otherworldly sense as forms of human media – technologies of reproduction rather than full-fledged sovereign (self-determining, self-owning) individual subjects."[45] We might also locate the practices of familial debt bondage in Whitehorse within this economy of self-lending. While there is nothing inherently progressive about this, it opens other ontologies of self, other social practices, other social values and other social logics that hold possibility for

Figure 4.6 The grand finale.

unanticipated histories and plural normative horizons. They might "set the stage," Tadiar speculates, "for radical departure from the given conditions of life under empire now."[46]

Tlingipino Bingo possibly set the stage for other narratives about Filipinos in Yukon, beyond those of successful nannies and hard-working Filipinos, and other models of inclusion beyond liberal citizenship. This likely would entail recognising in Gramma Susie's welcome claims to territory and agency that emerge from within traditions of hospitality somewhat different than state multicultural inclusion.[47] Gramma Susie's welcome might be viewed as a call to relationship that sets in train expectations not only of reciprocity but responsibility and responsiveness to the agency and concerns of First Nations peoples and the land in which they have settled.[48]

Although not named in the production, *Tlingipino Bingo* may have generated some new possibilities for relationship with settlers as well. Consider, for example, the exchange that we had a couple of days after the event with an older white Elks Lodge volunteer, who has run bingo for the past 40 years and espouses some conservative policy views – such as drug testing of Indigenous peoples in order for them to access social assistance benefits. Unprompted, he said, "I loved your performer." He persisted with genuine curiosity: "What do you call him?" [A drag queen.] "Is his lifestyle that or is he just performing?" [Yes, he's gay.] "But he doesn't dress like that everyday?" He concluded, "Yeah he's kinda neat. One of our bartenders was out and having a cigarette and he was out there. He told her that you guys were probably going bar hopping or something [after the show]. He said he had to get back to the hotel because he had this cute little black dress to wear." "He's kinda neat."[49]

Unsettled

During the intermission, a white audience member came up to one of us and said with great warmth and enthusiasm that *Tlingipino Bingo* was the most multicultural space she had ever been in during all of her years living in Whitehorse. Sharon Shorty's story – delivered in the third round of bingo, after she stepped out of her Gramma Susie outfit and character, and returned to enjoy the game as herself – could be interpreted within this framework of multiculturalism as well. One of the prizes for what was called the Tlingipino round of bingo games was a bouquet of barbequed pork skewers prepared by the CFAY. CFAY barbequed skewers are well known in Whitehorse because CFAY prepares hundreds of them to be sold at the Whitehorse Canada Day celebration each year. Their booth is, we were told, one of the most popular, and line-ups to purchase the skewers are long. After the prize of skewers was won, Sharon shared this story:

> Here's a story about these wonderful pork skewers. Does anybody else like pork skewers on July 1st? [Cheers.] From the Yukon Filipino Association. Thank you. So, when I was dating my husband, Derek – Derek Yap stand up – when we were dating, I told him, 'Please don't waste money on flowers. I can't eat flowers. Bring me meat on a stick. I'm an Indian woman.' So, on July 1st, on our anniversary, he comes running

> with 6 skewers from the Filipino Association, saying 'Happy Anniversary!' And I was like, 'I love you.' It's true love. And now he always brings me meat on a stick.

While this could be interpreted as instantiating a quintessential state multicultural moment (sharing "ethnic" food at a Canada Day celebration), exchanges across food (and the event of *Tlingipino Bingo*) might be more complex. They might unsettle rather than solidify the settler-colonial nation. A number of Filipinos with whom we spoke began to trace the foundations for a relationship with Indigenous peoples in shared food traditions: "We love to eat. Wild meat"; "[There are] quite a few similarities. In terms of the food, they like fish heads, we love fish heads." Particularly with the latter observation, commonality is implicitly constructed against the recognition that many white Canadians would find fish heads repugnant.

White settlers, with the exception of artists, were allowed to participate and witness but were not fully welcomed into the *Tlingipino Bingo* game, and resentment built when the same white person won the first two games. The director of *Tlingipino Bingo* observed, "And they [the white audience member] kept on winning. What does that mean? Twice. [...] What is this? This is a game of chance, yes. But it's not. Twice over. And he was jumping [with delight]. He was jumping [when his number was called]!" It is fair to say that we, as participants in the creation and production of *Tlingipino Bingo*, and acting under the dubious sign of university researchers (from the south of Canada), never escaped suspicion ourselves, never found a place in the process in which to settle comfortably. Nor should we. How then can we, as white settlers, be accountable and responsive to the gift of *Tlingipino Bingo*, a performance emerging from *Nanay*?[50]

In orienting towards decolonisation – a term, as white settler, Emilie Cameron uses with caution and without claiming for herself – she urges non-Indigenous scholars to learn not only the shape of our influence and claims but also the limits of those claims, and all the ways that we do not matter and do not know. "We must learn," she writes, "to know less, claim less, to listen, and to stop."[51] For us in the process of helping to create *Tlingipino Bingo*, and for all of the non-Indigenous and non-Filipinos in the Elks Lodge, the event was a lesson in listening and sometimes taking the brunt of the mockery and resentment. It was a lesson in engaging with the possibility that our presence was unwanted or at best irrelevant to that event. We need, in Cameron's wording, "to continually articulate methodological and theoretical spaces within which both the reach and limits of [settler] practices might be apprehended."[52] It is possible that *Tlingipino Bingo* was such a space.

Writing about *Tlingipino Bingo* opens another opportunity to think about and question the ongoing violence of settler colonialism in Yukon. It provides a critical space to contemplate the possibility that Filipino immigrants – without malice or intent – occupy a role within settler colonialism as a model minority discourse travels beyond the Filipino community. But are we the ones to make such a critique or is this critique another repetition of colonial violence?[53] We understand our responsibility as one of understanding how the discourse of model minority circulates within a wider Canadian society; our critique is of the model minority discourse and not of the Filipino community. We write about *Tlingipino Bingo* so that we might better understand its political possibilities, as one means of "setting the stage" to imagine relationship beyond citizenship. Alongside narratives of successful nannies and hard-working model minorities, it seems to offer other narratives to the Filipino community in Whitehorse. These are narratives that add a layer of complexity to the stories of labour exploitation and sadness that we have circulated through *Nanay*, which did not resonate or have political traction within the Filipino community there.

Nanay's lack of political traction in Whitehorse gives reason to pause. On the one hand, it brings to light an important limitation in migration studies: this is that most theorising about migration tends to emerge from experiences within a few larger cities. This tendency is another instance or a different face of methodological nationalism whereby the variety and subtlety of social existence is collapsed within a presumed homogeneous national time-space.[54] Indeed, processes and patterns of Filipino migration to Whitehorse have been different from those in Vancouver (where *Nanay* was created), and Filipinos are brought into very different relationships with white settler society and Indigenous communities in these two places. *Nanay* inadvertently brought us to a comparative analysis of Filipino migration to Canada. At the same time, the fast incorporation in Whitehorse of the Filipino migrants as reliable model-minority service-sector workers, many in precarious dead-end jobs, invites a critical analysis of the discursive workings of settler colonialism. The humour and companionability of *Tlingipino Bingo*, merged with a sharp critique of ongoing settler colonialism and the possibilities of different modes of less hierarchical relations, with other communities, the land and other beings in Yukon, offers one such political possibility. *Tlingipino Bingo* may open other narratives and critical political possibilities for claiming a space in Canada in solidarity with Indigenous peoples, possibly less imbricated within settler colonialism.

Finally, we put stories of *Tlingipino Bingo* into circulation to express our gratitude for this gift and as a way – possibly – of extending relations of accountability through its circulation.[55] We accept that writing about *Tlingipino Bingo* may simultaneously instantiate settler colonialism in familiar ways, as we – white settlers – impose an interpretation and then translate and transport the performance into the academic archive and a particular form of knowledge production. Whether our writing is an act of extraction (the act of taking a process out of the relationships that give it meaning) or abstraction (shifting relationality to illuminate another meaning),[56] we are uncertain. This undecidability about writing about *Tlingipino Bingo* keeps us attuned to the perniciousness of settler colonialism, and – we hope – nurtures the possibility for further conversation within the Indigenous-arrivant-settler triad.

Notes

1 See, for example, Jodi Byrd, *Transit of Empire: Indigenous Critiques of Colonialism* (Minneapolis, London: University of Minnesota Press, 2011); May Chazon, Lisa Helps, Anna Stanley, and Sonali Thakkar, eds., *Home and Native Land: Unsettling Multiculturalism in Canada* (Toronto: Between The Lines, 2011); Iyko Day, *Alien Capital: Asian Racialization and the Logic of Settler Colonial Capitalism* (Durham and London: Duke University Press, 2016); May Farrales, "Gendered Sexualities in Migration: Play, Pageantry, and the Politics of Performing Filipino-ness in Settler Colonial Canada" (unpublished PhD dissertation, University of British Columbia, 2017); Bonita Lawrence and Enakshi Dua, "Decolonizing Antiracism," *Social Justice* 32, no. 4 (2005): 120–143; Lisa Lowe, *The Intimacies of Four Continents* (Durham and London: Duke University Press, 2015); Dean Saranillio, "Why Asian Settler Colonialism Matters: A Thought Piece on Critiques, Debates, and Indigenous Difference," *Settler Colonial Studies* 3 (2013): 280–294; Audra Simpson, *Mohawk Interruptus: Political Life across the Borders of Settler States* (Durham and London: Duke University Press, 2014); Anna Stanley, Sedef Arat-Koc, Laurie Bertram, and Hayden King, "Intervention – Addressing the Indigenous-Immigration 'Parallax Gap,'" *Antipode Foundation* (blog), 18 June 2014; Neferti X.M. Tadiar, "Decolonization, 'Race,' and Remaindered Life Under Empire," *Qui Parle: Critical Humanities and Social Sciences* 23, no. 2 (2015): 135–160; Kim TallBear, *Making Love and Relations Beyond Settler Sexualities* (Uppsala: Technoscience Research Platform, 2016), Video recording; Rita Wong, "Decolonization: Reading Asian and First Nations Relations in Literature," *Canadian Literature* 199 (2008): 158–180.

2 Our intention was to support the project through initial funding and grant writing and to document the production that came out of this. As the project evolved, for a variety of work-related, health and personal reasons, we took on a more central role creating the structure of the show in collaboration with others, and as co-producers and co-curators of the performance.

3 Statistics Canada, "2016 Census: City of Whitehorse, Yukon and Yukon Territory," Census Profile, Ottawa, 2017. The city of Whitehorse exists on the traditional territories of the Kwanlin Dün and Ta'an Kwäch'än peoples.

4 For play development and research, we made four trips to Whitehorse, which ranged from three days to two weeks. Interviews were carried out with fifteen members of the Filipino community in Whitehorse, as well as two government representatives, and a member of the Whitehorse Elks. Filipino interviewees were asked about their migration histories and to reflect on their lives and the Filipino community in Whitehorse.

5 As fate would have it, Teslin is the seat of Tlingit self-government.

6 In recent years, the programme has become more closely regulated: the government has added a language requirement, contacts the prospective employer and more closely scrutinises the suitability of work experience: "It's not like 'I work as a scuba diving instructor and now I'm going to work as a food and counter attendant at McDonald's.' You need to demonstrate that you worked for the food industry for a minimum of 6 months. [...] Before we kind of took your word for it, now we actually contact employers."

7 Ayse Çaglar and Nina Glick Schiller, *Migrants and City-Making: Dispossession, Displacement, and Urban Regeneration* (Durham and London: Duke University Press, 2018); Nina Glick Schiller and Ayse Çaglar, "Migrant Incorporation and City Scale: Towards a Theory of Locality in Migration Studies," *Journal of Ethnic and Migration Studies* 35, no. 2 (2009): 177–202.

8 Genesee Keevil, "The Philippine Connection: How One Woman Turned Whitehorse into a Promised Land," *Walrus*, 18 May 2016.

9 Lisa Davidson, "(Res)sentiment and Practices of Hope: The Labours of Filipina Live-In Caregivers in Filipino Canadian Families," in *Filipinos in Canada: Disturbing Invisibility*, eds. Roland Sintos Coloma, Bonnie McElhinny, Ethel Tungohan, John Paul Catungal, and Lisa M. Davidson (Toronto: University of Toronto Press, 2012), 142–160.

10 Eve Tuck and K. Wayne Yang, "Decolonization Is Not a Metaphor," *Decolonization: Indigeneity, Education & Society* 1, no. 1 (2012): 1–40.

11 Jodi Byrd, *Transit of Empire: Indigenous Critiques of Colonialism* (Minneapolis, London: University of Minnesota Press, 2011).

12 Lisa Lowe, *The Intimacies of Four Continents* (Durham and London: Duke University Press, 2015).

13 Jodi Byrd, *Transit of Empire: Indigenous Critiques of Colonialism* (Minneapolis, London: University of Minnesota Press, 2011), xix.

14 Eva-Lotta E., Hedman, and John T. Sidel, *Philippine Politics and Society in the Twentieth Century: Colonial Legacies, Post-colonial Trajectories* (London: Routledge, 2000), 38; see also Stuart Creighton Miller, *Benevolent Assimilation: The American Conquest of the Philippines, 1899–1903* (New Haven: Yale University Press, 1982).

15 Neferti X.M. Tadiar, "Decolonization, 'Race,' and Remaindered Life Under Empire," *Qui Parle: Critical Humanities and Social Sciences* 23, no. 2 (2015): 143.

16 Ibid., 143.

17 Patrick Wolfe, "Settler Colonialism and the Elimination of the Native," *Journal of Genocide Research* 8, no. 4 (2006): 387–409.

18 Dean Saranillio, "Why Asian Settler Colonialism Matters: A Thought Piece on Critiques, Debates, and Indigenous Difference," *Settler Colonial Studies* 3 (2013): 286; see also Laura Pulido, "Geographies of Race and Ethnicity III: Settler Colonialism and Non Native People of Color," *Progress in Human Geography* 42, no. 2 (2018): 309–318.

19 We thank Emilie Cameron for this wording.

20 For example, Krystle Alarcon, "First Filipina nanny helped community grow," *Yukon News*, 30 August 2013; Joshua Clarkson, "Foreign workers take jobs that nobody else wants," *Yukon News*, 11 March 2015; Raymond De Souza, "The Yukon is the north, pure and simple," *The National Post*, 7 June 2012; Tristin Hopper, "Foreign workers championed despite slowdown," *Yukon News*, 3 April 2009; Chris Oke, "Yukon retains its foreign workers," *Yukon News*, 8 June 2011; Jacqueline Ronson, "Film fest brings world stories home," *Yukon News*, 1 February 2013; John Thompson, "Territory to expand foreign worker program," *Yukon News*, 18 June 2010; Werner Walcher, *Cold Paradise* (Yukon: Cold Paradise Productions, 2014), Video recording; Josh Wingrove, "Filipino community thrives in Yukon; Streamlined path to entry, residency appeals to immigrants who come to work," *Globe and Mail*, 23 January 2014, A4.

21 Iyko Day, *Alien Capital: Asian Racialization and the Logic of Settler Colonial Capitalism* (Durham and London: Duke University Press, 2016).

22 Bonnie Honig, "Immigrant America? How Foreignness 'Solves' Democracy's Problems," *Social Text* 56 (1998): 1–27; David Pulumbo-Liu, *Asian/American: Historical Crossings of a Racial Frontier* (Stanford: Stanford University Press, 1999). In the U.S. context, Pulumbo-Liu documents the ways that white supremacist ideology has been channelled through the body of the Asian American, in particular, its capacity for hard work, perseverance and independence from state or federal largesse. Byrd argues that the Asian body now bears and takes forward the "cowboys and indians" narrative in the United States. To sustain this point, she draws on Pulumbo-Liu's analysis of popular media representations of Korean Americans during the Los Angeles riots in 1992 as vigilante "cowboys" protecting their property from looting by African Americans.

23 We thank Emilie Cameron for this phrasing.

24 See www.sharonshorty.com/historysc.htm.

25 Reflecting on his celebrated theatre play, *The Rez Sisters*, which revolves around the journeys of seven bingo-obsessed Indigenous women in Southern Ontario after they hear rumour that the world's biggest bingo game is coming to Toronto, Tomson Highway has observed, "I'm sure some people went to *Rez* expecting crying and moaning and plenty of misery, reflecting everything they've heard about or witnessed on reserves. They must have been surprised. All that humour and love and optimism, plus the positive values taught by Indian mythology." (quoted in Denis Johnston, "Lines and Circles: The 'Rez' Plays of Tomson Highway" Canadian Literature 124–25 (1990): 259). Thomas King, the celebrated American-Canadian Indigenous writer and humourist, speaking about his work on the Canadian radio series *Dead Dog Café*, explains how he turned to humour as a "survival strategy… If life is so bad, you either kill yourself or you laugh. Colonized people can see humour as a strength, as a medicine" (quoted in John Stackhouse, "Comic heroes or 'red niggers'?" *Globe and Mail*, 9 November 2001). "I'm dealing with cultural humour," he says, "And occasionally, I'm going to hold that mirror up to white Canada, and say, 'See this is what it feels like'. […] You can get into the back door with humour. You can get into their kitchen with humour. If you're pounding on the front door, they won't let you in" (quoted in Drew Hayden Taylor, *Redskins, Tricksters and Puppy Stew* (Montreal: National Film Board of Canada, 2000)).

26 Scott Lauria Morgensen, "Unsettling Queer Politics: What Can Non-natives Learn from Two-Spirit Organizing?" in *Queer Indigenous Studies: Critical Interventions in Theory, Politics, and Literature*, eds. Qwo-Li Driskoll, Chris Finley, Brian Joseph Killey, and Scott Lauria Morgensen (Tucson: University of Arizona Press, 2011), 132–154.

27 Sarah Hunt and Cindy Holmes, "Everyday Decolonization: Living a Decolonizing Queer Politics," *Journal of Lesbian Studies* 19 (2015): 159.

28 Martin F. Manalanson IV, *Global Divas: Filipino Gay Men in Diaspora* (Durham and London: Duke University Press, 2003).

29 Ibid., 42, 140.

30 Fenella Cannell, *Power and Intimacy in the Christian Philippines* (Cambridge: Cambridge University Press, 1999), 223.

31 Kent Monkman quoted in Robert Everett-Green, "Kent Monkman: A Trickster with a Cause Crashes Canada's 150th Birthday Party," *Globe and Mail*, 7 January 2017.

32 Filipino immigration to Canada is understudied more generally. As a corrective to this tendency, see Coloma et al. (2012).

33 Maya Mikdashi, "What Is Settler Colonialism? (For Leo Delano Ames, Jr.)," *American Indian Culture and Research Journal* 37, no. 2 (2013): 32.

34 See Sarah Hunt and Cindy Holmes, "Everyday Decolonization: Living a Decolonizing Queer Politics," *Journal of Lesbian Studies* 19 (2015); Audra Simpson, *Mohawk Interruptus: Political Life across the Borders of Settler States* (Durham and London: Duke University Press, 2014).

35 Hannah Arendt, "Introduction: Walter Benjamin: 1892–1940," in *Illuminations* by Walter Benjamin, trans. Harry Zohn (New York: Schocken Books, 1969), 1–55.

36 Coll Thrush, *Indigenous London: Native Travelers at the Heart of Empire* (New Haven: Yale University Press, 2016).

37 Audra Simpson, *Mohawk Interruptus: Political Life across the Borders of Settler States* (Durham and London: Duke University Press, 2014).

38 Philip J. Deloria, *Indians in Unexpected Places* (Lawrence: University of Kansas Press, 2004); Audra Simpson, *Mohawk Interruptus: Political Life across the Borders of Settler States* (Durham and London: Duke University Press, 2014).

39 Floro Mercene, *Manila Men in the New World: Filipino Migration to Mexico and the Americas from the Sixteenth Century* (Quezon: University of the Philippines Press, 2007).
40 Renisa Mawani, *Colonial Proximities: Crossracial Encounters and Juridical Truths in British Columbia 1871–1921* (Vancouver: University of British Columbia Press, 2009).
41 Sedef Arat-Koç, "An Anti-Colonial Politics of Place," in *Intervention – Addressing the Indigenous-Immigration 'Parallax Gap,'* eds. Anna Stanley, Sedef Arat-Koc, Laurie Bertram, and Hayden King, *Antipode Foundation* (blog), 18 June 2014: n.p.
42 Dipesh Chakarbarty, *Provincializing Europe: Postcolonial Thought and Historical Difference* (Princeton: Princeton University Press, 2000): 20.
43 See also Kim TallBear, *Making Love and Relations beyond Settler Sexualities* (Uppsala: Technoscience Research Platform, 2016), Video recording.
44 Neferti X.M. Tadiar, "Decolonization, 'Race,' and Remaindered Life Under Empire," *Qui Parle: Critical Humanities and Social Sciences* 23, no. 2 (2015): 147.
45 Ibid., 152.
46 Ibid., 156.
47 For a parallel discussion in the Australian and New Zealand contexts, see Emma Cox, "Sovereign Ontologies in Australia and Aotearoa-New Zealand: Indigenous Responses to Asylum Seekers, Refugees and Overstayers," in *Knowing Differently: The Challenges of the Indigenous*, eds. G.N. Devy, Geoffrey Davis, and K.K. Chakravarty (New Delhi: Routledge, 2013), 139–157; Rand Hazou,"Performing Manaaki and New Zealand Refugee Theatre," *Research in Drama Education* 23, no. 2 (2018): 228–241.
48 Rauna Kuokkanen, *Reshaping the University: Responsibility, Indigenous Epistemes, and the Logic of the Gift* (Vancouver: University of British Columbia Press, 2007).
49 For the significance of daily actions in everyday spaces for decolonial processes and practices, see Sarah Hunt and Cindy Holmes, "Everyday Decolonization: Living a Decolonizing Queer Politics," *Journal of Lesbian Studies* 19 (2015).
50 Another approach would be to respond seriously to the questioning we received about our own ancestry and to situate ourselves more precisely within the historical narratives we seek to disturb. Analysing the U.S. media's focus on and representation of Korean American and African American conflict during Los Angeles Riots of 1992, Pulumbo-Liu notes that it left white Americans "free both to stand apart and speculate" on the conflict (David Pulumbo-Liu, *Asian/American: Historical Crossings of a Racial Frontier* (Palo Alto: Stanford University Press, 1999), 193). Witnessing settler colonialism and conflict between Filipinos and First Nations without implication seems to repeat this dynamic. Just a little research into our family histories reveals historical complicities that are more immediate and more intimate than we had imagined, and our ignorance of these histories is itself politically significant. Caleb learned more concerning the move of his grandparents to Whitehorse with their four children in 1964, driving north from Vancouver on a gravel road to settle in purpose-built government housing designed for the civil service and army. His grandfather had taken a job with the Ministry of Indian Affairs, working as a counsellor in the territory's residential schools. He was – by lineage – implicated in aggressive state policy designed to enact the violent assimilation and/or cultural annihilation of Indigenous peoples in the Yukon, implicated in a horrendous "civilising" strategy that forcibly removed Indigenous children from their family homes, stripping them of their ancestral languages, exposing many to horrific abuse. We rented accommodation in the same housing development during our stay in Whitehorse. Gerry's great great grandmother, Valdis

Gudmundsdottir, was among the first 365 Icelandic settlers to arrive in New Iceland, on the shores of Lake Winnipeg in Manitoba in 1875, three days after the Dominion Order-in-Council granted the Icelanders their 600-square mile block settlement reserve. This land was by no means vacant or uncontested and was in fact granted before aboriginal title was extinguished (Anne Brydon, "Dreams and Claims: Icelandic-Aboriginal Interactions in the Manitoba Interlake." *Journal of Canadian Studies* 36, no. 2 (2001): 164–190). Granting inclusion to European migrants while simultaneously suspending the rights of another community as part of a larger state effort to eliminate the latter is, as Laurie Bertram observes, "the early expression of sovereign-power through the creation of race-based states of exception, or the erosion or suspension of the rights of a certain population as part of a larger attempt to eliminate that population" (Laurie Bertram, "Resurfacing Landscapes of Trauma: Multiculturalism, Cemeteries, and the Migrant Body, 1875 Onwards," in *Home and Native Land: Unsettling Multiculturalism in Canada*, eds. May Chazon, Lisa Helps, Anna Stanley, and Sonali Thakkar (Toronto: Between The Lines, 2011), 157–174). The Icelanders were part of a state-coordinated population transformation known as the "peopling of the prairies" (Ibid., 162). As a talented midwife, Valdis contributed to this biopolitical project by midwifing the birth of the first Icelander born in Canada. On the other side of biopolitics, Icelanders brought smallpox that killed 10%of the Icelanders and 70% of the local Indigenous population (Ibid., 162). Valdis and her husband Simon Simonarson lived intimately with these deaths, taking into their home a number of children whose Icelandic parents had died of smallpox. Anglo-Canadians termed Icelanders "squawmuck" and characterised them as a "seal skin wearing and blubber eating race." Icelanders worked to whiten themselves in their early years in Manitoba, and in an effort to reform the community's image, the Icelandic elite in the 1880s initiated a campaign to support the government's war against Indigenous land claims and to suppress the Northwest Rebellion (Laurie Bertram, "Icelandic and Indigenous Exchange, Overlapping Histories of Migration and Colonialism," in *Intervention – Addressing the Indigenous-Immigration 'Parallax Gap,'* eds. Anna Stanley, Sedef Arat-Koc, Laurie Bertram, and Hayden King. *Antipode Foundation* (blog), 18 June 2014, n.p.).

We have wrestled with whether our histories should be in the text or endnote. Our point in pursuing our histories is not be to recentre whiteness or to seek a space of reconciliation within *Tlingipino Bingo*; it is to say, yes, we carry family histories that put us into this picture as well, and intimately so, and to say that colonialism is an active and ongoing process, including the process of becoming white. Writing these histories into our account of *Tlingipino Bingo* could be seen as one effort to disrupt "the habit among settlers of imagining that we are somehow not involved" (Emilie Cameron, *Far Off Metal River: Inuit Lands, Settler Stories, and the Making of the Contemporary Arctic* (Vancouver: University of British Columbia Press, 2015), 17). Now more than ever, claiming such histories could be a means to simultaneously assert and unsettle our claims to nation space in an effort to disrupt the nostalgia for a lost time of national whiteness. To un(document) oneself or to trace the arbitrary and tenuous legal nature of one's inclusion is possibly an act of solidarity with those who live their lives without full benefits of citizenship. Still, in the context of *Tlingipino Bingo*, the specificity of our history did not alter our participation. As such, we may be more appropriately left to embody with discomfort our assigned roles as representatives of a structure of privilege. We leave our histories in the margins. We thank Emilie Cameron for this point. We thank Sarah Hunt and Cindy Holmes for their clear articulation of the distinction between a politics of accountability and a politics of inclusion in relations of allyship. See Sarah Hunt and Cindy Holmes, "Everyday Decolonization: Living a Decolonizing Queer Politics," *Journal of Lesbian Studies* 19 (2015): 154–172.

51 Emilie Cameron, *Far Off Metal River: Inuit Lands, Settler Stories, and the Making of the Contemporary Arctic* (Vancouver: University of British Columbia Press, 2015), 14, 20.
52 Ibid., 23.
53 We note that it is one that is being actively explored by Filipino activists (e.g., May Farrales, "Gendered Sexualities in Migration: Play, Pageantry, and the Politics of Performing Filipino-ness in Settler Colonial Canada" (unpublished PhD dissertation, University of British Columbia, 2017)), including a minority we met in Whitehorse. Given the size of the community in Whitehorse and protocols of confidentiality, we have not identified these individuals.
54 Nina Glick Schiller and Ayse Çaglar, "Migrant Incorporation and City Scale: Towards a Theory of Locality in Migration Studies," *Journal of Ethnic and Migration Studies* 35, no. 2 (2009): 177–202.
55 Rauna Kuokkanen, *Reshaping the University: Responsibility, Indigenous Epistemes, and the Logic of the Gift* (Vancouver: University of British Columbia Press, 2007); see also Pat Noxolo, Parvati Raghuram, and Clare Madge, "Unsettling Responsibility: Postcolonial Interventions," *Transactions of the Institute of British Geographers* 27, no. 3 (2012): 418–429.
56 Audra Simpson, *Mohawk Interruptus: Political Life across the Borders of Settler States* (Durham and London: Duke University Press, 2014), 202.

Conclusion

Relations, refusals and openings

Nanay grew out of a frustration with a kind of comparative analysis that feeds Canadian complacency with – and complicity in – a process of migration that causes great hardship and struggle for many Filipino families. It is a complacency that equally forestalls serious engagement with a collective responsibility for childcare and eldercare in Canada. It is estimated that 44% of non-school-aged children in Canada live in childcare deserts, that is, where there are more than three children for every available licensed space and hence childcare is difficult or impossible to access.[1] Canada is at the bottom (32 of 38) of Organisation for Economic Co-operation and Development (OECD) countries for public spending on services for families and children.[2] Responding to the question posed by the OECD, A Good Life in Old Age?, the answer in Canada would appear to be "No," if long-term care expenditure as a share of gross domestic product is any measure: expenditure in Canada is well below the OECD average.[3] Since 2001, in BC, the provincial government has cut health and social services substantially, closed over 50 healthcare facilities including long-term care homes and hospitals, prioritised for-profit long-term care delivery, narrowed service entitlement and devolved eldercare provision onto families, provoking what many term a crisis in long-term (and elder) care.[4] Canada imports poor "third world" women to provide care because there has been little societal commitment to collectively bear its costs. Bringing individual skilled workers to do socially necessary work and requiring that they leave their families at home to be cared for in the Philippines while they do it, Canadians create the conditions that result in the deskilling of university-educated women, and the estrangement of families after long periods of separation. The costs of the social reproduction to domestic workers' families are passed onto their families in the Philippines, in a context where there is also minimal governmental support. And yet the LCP is typically framed within a heroic national narrative, of Canadians saving Filipinos from the dire poverty of their homeland through the opportunity to migrate to Canada after fulfilling a limited number of years as precarious migrant workers. This heroic narrative[5] depends on a comparison between Canada and the Philippines, as well as a comparison with other countries to which Filipinos migrate as temporary foreign workers; the capacity to migrate permanently to Canada after a type of indentured servitude often casts Canada in a favourable light relative to other countries receiving and utilising this labour. Changes to the programme since 2014 suggest that Canadians have even less claim to this comparative advantage.[6] We put *Nanay* in the middle of Canadians' unwarranted complacency to try to shake it up and spark a simultaneously impassioned and informed public debate.

Relational comparison

Another kind of relational comparative analysis and politics emerged as the play took us on its travels, and as we were able to embed processes in Canada in a larger circuit of neo-liberal and colonial relations and effects. Certainly, this more expansive analysis is possible without a travelling play. Robyn Rodriguez, for instance, traces the almost imperceptible but nonetheless persistent filaments of U.S. colonial influence that replicate and expand the dispersal of Filipino migrant labour around the world: one example is the significance of U.S. corporations for the employment of OFWs in Saudi Arabia. Such examples highlight, she argues, the "deep linkages" between the Philippine labour brokerage state, the globalisation of U.S. capital and the Philippine labour diaspora that exist both indirectly and directly throughout the world.[7] For us, the play as a thing took us on a tour of relational comparisons: it *relationed* comparisons. The view from the Philippines via the play made the full impact of the LCP more readily apparent and deeply felt: a university student being comforted by his friends as he spoke in a post-performance public forum of the plenitude of his life in Manila and his deep desire not to join his mother in Canada; a young child crying in the street after seeing the play in Bagong Barrio because he missed his mother who was working in Canada under the LCP; middle-class medical students and marginalised youth alike emphatically refusing the invidious comparison between the Philippines and Canada and the inevitability of a life as an OFW. In the Philippines, Canada's place at the peak of the migration hierarchy was both named and questioned. Stories of workers literally used up and discarded in old age through lifetimes of work as temporary migrant workers in Saudi Arabia confirmed it, but the conscious decision to take temporary positions working as a nurse in Saudi Arabia over the prospect of being deskilled as a caregiver in the LCP in Canada upended it.[8] We could see the ruthless working of neo-liberal capitalism on the ground in Bagong Barrio, as hard-working, fiscally responsible, future-oriented households planned and plotted to insert themselves into the migration process in order to survive. Canada is not exceptional; it is one cog in a massive migration machine. In a single family, one sibling might go to Canada, another work on a ship as a seafarer, another set off for Saudi Arabia. The same forces drive migration to Canada and elsewhere and domestic workers' experiences of abuse and humiliation reverberate and resonate from Kuwait to Vancouver.

The view from Whitehorse in the Canadian North opened another set of relationships, including the repetition of different experiences of family separation within settler colonialism. The state violence of family separation has been enacted within Indigenous communities as a form of cultural extinction solidifying dispossession from the land, and within Filipino families as a means of cutting the costs of care-work, extracting the labour of domestic workers while abstracting it from their social relations. The play brought us to the interface of colonialism, settler colonialism and racial capitalism, the latter we understand to be the use of racial distinction to derive value and as a structuring analytic in capitalist societies. The weight of our analysis is that Filipino immigrants are not immune to the dynamic set in place within settler colonialism but resonating histories of colonial violence open room for imagining other possibilities.

Contingent politics

What possibilities? Some audience members were irritated that we did not offer clearer directives for politics and action. As one cranky audience member wrote on the feedback questionnaire at a Vancouver performance, "The pretentiousness of the people who are going 'Oh no, the poor third world women are having their human rights violated.' The sentiments are wonderful but give them something to act on." At the PETA Center performance, another argued that testimonial theatre demands a more explicit argument: "testimonies are used to present an argument, whether it's going to be used for the prosecution or the defense." "What is the stand," we were asked, "regarding the very lives that get to be told here?" But we understood that the point of the play was to present complexity for discussion and a space for politics, rather than providing clear-cut answers.

The politics of Filipino migration took and takes shape differently in different contexts. The Vancouver performance was framed through feminist anti-racist commitments to social justice, inclusion and a fundamental revaluation of care and care-work. The most radical moments came when audience members called for a total revisioning of the distribution of work and care in households and communities. Other moments called for – or came to – a fuller understanding, through discussion, of the problems with a temporary worker programme that arbitrarily excludes workers' children from migration. The play and the forums that followed made visible and commonsensical the necessity for alliances between migrant organisations and advocates in the childcare and healthcare sectors: to fight for admission of domestic workers as permanent immigrants, rather than as probationary temporary foreign workers, on the understanding that they are doing skilled and highly valuable work; and to align over workers' rights across the childcare and healthcare sectors.

The risks or limitations of this sought-after inclusion – an unquestioned good in the Vancouver performances – were sounded in different registers in Manila and Whitehorse, from anti-neocolonial and decolonial perspectives.

From Manila, one concern is that the ideal of inclusion in Canada merely props up the continued reliance of the Philippine economy on labour export and remittances as a key development strategy, with the approval and encouragement of hegemonic economic organisations such as the World Bank.[9] The promise and prospect of Filipino inclusion in Canada potentially directs attention away from the conditions that force migration from the Philippines and divides interests within the Filipino diaspora. In short, the aspirations voiced and emerging after the Vancouver performances sit uneasily with important debates within and beyond the Philippines. Differently and equally, successful Filipino integration in Canada can (and, we have argued, does in some circumstances in Whitehorse) play into the ongoing processes of settler colonialism.

Although doubts about inclusion may seem misaligned with the LCP, given that the potential for citizenship is typically considered to be its most attractive feature, this is precisely the point of more careful consideration.[10] When audience members in the Philippines spoke against the culture of migration, of refusing their parents' offer to sponsor their migration to the United States or Canada, or their decision to go to Saudi Arabia as a nurse on a series of short-term contracts rather than enter the LCP: this is a robust politics of refusal. As much as Migrante International supports migrants in their forced choice of migration (precisely because it is understood to be forced), refusing labour export as a development strategy grounds their politics. In Canada, this politics of refusal exists in varying forms as well. There have been vigorous discussions in Canada among Filipino activists of how to position themselves in relation to the Philippine diaspora: from the perspective of Canada or the Philippines as home. Orientations to different homelands invite different alliances and solidarities. Some in the Filipino community in Canada have also taken up doubts about inclusion within Canada as a settler-colonial nation, as a way of imagining living in Canada differently.[11] We approach this politics with caution, however, fully conscious of our status as privileged white settlers whose Canadian citizenship is not in doubt. We approach with caution, insisting on the contingency and specificity of such politics, because many people in many countries in the global north and elsewhere are all too willing to engage in a reciprocal project of refusal.[12] Considering the LCP specifically, the programme has been restructured since 2014 such that it is no longer a de facto pathway to citizenship and it will be under further review in 2019.[13] Further, activist organisations report increased policing of domestic workers registered in the LCP under "Project Guardian" of the Canadian Border Services Agency (CBSA).[14] As access to citizenship becomes more precarious, critiques of inclusion need to be parsed very carefully.

Writing of the refusal of a Canadian passport by members of the Mohawk of Kahnawàke nation, Audra Simpson notes, there are no easy answers. In thinking about the possibilities of refusal that might align across migrant and Indigenous communities, she writes appreciatively of Gilberto Rosas' ethnography of marginalised Mexican youths who, living in Nogales at the border of Arizona and Sonora, call themselves Libre Barrio (Free 'Hood). Moving with some fluidity across the border, "Barrio Libre was more than a free-floating geography, superimposed over a dominant one [...] to belong to it was an expansive, furious refusal of normativity, an enraged subversion of the respective sovereignties of the U.S. and Mexico that seeped from under the new frontier."[15] That this refusal involved its own violent practices, such as preying on undocumented migrants moving through the sewer system to cross the border between Mexico and the United States, is neither condoned nor condemned by Rosas (or Simpson): it is understood as one response within the dynamics of militarised policing and neo-liberal governmentalities. Though Mohawk refusals of the U.S.-Canada border and Canadian state authority over their political membership differ from those explored by Rosas, Simpson sees in them a resonant politics of refusal: the refusal on the part of Barrio Libre youths "to see [their] condition as anything other than a state of freedom is a refusal for us of the easy answer, of a structure of consent, of ease." They "smash" the "categorical imperatives" of easy answers.[16]

What could refusal of the easy answers mean for the Filipino community in Canada? Can we imagine relations of consent to Canada for Filipino immigrants that are not structured by the logics of settler colonialism within the discourse of model minority? What are (or are there) political possibilities within or alongside Canadian citizenship? Carol McGranahan's analysis of the refusal of citizenship among Tibetan refugees is of interest in relation to these questions, in particular her understanding that refusal takes shape differently in different places (and likely at different times). Many Tibetans exiled in Nepal and India hold a passport-sized "Green Book" issued by their exiled government and they have refused citizenship in these countries so as to maintain their claims to Tibetan sovereignty. Their relationship to the governments and societies of Nepal and India is not one of inclusion through citizenship but rather interdependence through the offering and acceptance of refuge. This has been a viable strategy in India and Nepal because they have been able to stay on for generations as refugees, but not so in Canada or the United States. In the latter contexts, citizenship has been accepted as a morally (and the only) viable option.[17] The point of interest to us is that political strategy and whether or not citizenship is a viable or desirable option for people displaced from their country of origin depends on context. In places where citizenship is sought, there may also be room for the quieter kinds of refusal described by Erica Weiss, which take shape as abstention from some forms of engagement and other performative tactics rather than overt resistance.[18] Through *Tlingipino Bingo*, we suggest that political possibility might be discovered through comparative analysis across different yet resonant experiences of colonial violence. These affinities may suggest modes of relating and interdependency that exist within, alongside and beyond citizenship. Turning towards relations of care and reciprocity and away from white settlers during the process and performance of *Tlingipino Bingo* seemed to be an opening to something new, underlining the point that an answer to the question of political possibility for Filipinos in Canada is not for us to find. What we *can* offer to this ongoing conversation are observations from the vantage of *Nanay* and *Tlingipino Bingo*, and strategies (that we have attempted to deploy in this book) of dislodging universalising unsituated claims (including our own), with an eye to disrupting the habit of imagining that "somehow we are not involved."[19] We have aimed to support efforts to unsettle white settlers' claims to originary status in Canadian territorial space so as to open political space for more equitable and mutually accountable ways of relating within and beyond the Canadian nation.

Nanay and ecologies of knowledge production

Learning from *Nanay* was not simply analytical. It has been embodied. It has been felt within a process of creating relationships and exchanging stories. The play has been a thing that has gathered and extended worlds within and around it. From fieldnotes the first day of rehearsal at the PETA Theater Center in Manila when those working on the play met together for the first time, "Marichu introduced herself first. She started with professional interests but quickly moved to telling us about being a single mom. She then revealed her new passion for yoga. Alex (the director) quickly interjected that this is a passion shared by several others in the room." This process of creating small bonds within the cast and crew kept happening through jokes and gestures: "Patrick and Alex revealed that they were in a play recently in which they wore only Speedos.[20] DK told us that his father is poet laureate of Saskatchewan." When Alex asked for an update on the LCP, the theme was raised of possible exploitation within Filipino families sponsoring their own family members to Canada to work for them as live-in caregivers. Again, from notes, "Marichu spoke of the notion of *utang ng loob* (and May picked up on this too): the debt of gratitude. When Marichu began to talk, we couldn't determine if we had affronted her by applying the language of exploitation to describe relations within the Filipino family. But as she spoke it seemed that she was saying something more subtle: that *utang ng loob* is a powerful concept within Filipino psychology that has to be at once respected, and questioned and examined." As we began to work with the script, "There is something interesting about watching the Filipino actors work with the material, the kinds of information they want about the characters, for instance, the specifics of the region from which they originate. It makes a difference in terms of dialect and how the character is interpreted. All of this was lost in the Vancouver production." Speedo bathing suits (laughter), yoga, family, reading the body for meaning, locating previously unlocated narratives, long rehearsals and nervous openings, translating and discovering gulfs of ignorance and threads of connection: these too were always an integral and generative part of the research/performance process. Never entirely confident of our success, we have nonetheless aspired to what Richa Nagar has termed "hungry" translations: "a dynamic, multidirectional process of ethical and politically aware mediation among otherwise impermeable local diversities."[21]

These embodied socially dense translations extended beyond the small world of creating the play, opening up networks of relations and encounters. In Vancouver, the play created (paid) internships for Filipino activists and connected the PWC of BC to cultural funders and presenters. Performances brought child care advocates into direct conversation with migrant worker organisations, employers with domestic workers and family members, and researchers with artists. Migrante in Vancouver introduced us to Migrante International in Manila; the PWC of BC to their aligned groups in Manila, and the PETA production would bring these groups (who otherwise are not likely in conversation) into the same shared talkback space. Actor and producer Hazel Venzon's associations with the CFAY brought them to the play to meet Sharon Shorty, and into the Elks Lodge for a night of story sharing and a performed game of chance. A priest attending *Nanay* in Bagong Barrio brought the play to his neighbouring community. A Migrante organiser emerged from the project recommitted to mobilising theatre elsewhere in the Philippines as an organising strategy. *Nanay* was itself a means of puncturing seemingly "impermeable local diversities" to bring new conversations and new knowledge sharing into being. These openings are indeterminate in their effects and temporality. An unforeseen message brought the project to Berlin. Or, without any intermittent communication, one of us recently received this email, five years after the PETA production: "I'm hoping you remember me from when you were in Manila for the staging of *Nanay*. Was wondering if…." Like Emilie Cameron,[22] we are reluctant to claim a decolonial process for ourselves and our scholarship, and equally feminist aspirations of less hierarchical, mutually consensual, long-term, relational, reciprocally enriching knowledge practices are ones we continually reach towards (rather than necessarily achieve).[23] It may be our failures or tentative approximations, in fact, that bring us closer to our goal, in the sense that our positionality, authority and capacity to deliver on these ideals are always in doubt and always open to revision. What we can say without reservation is that we have valued opportunities to learn and that the play, as a worldly thing, *relationed*[24] into being multidirectional communication between local diversities and situated solidarities. It spawned a complex network of embodied relations within a rich ecology of knowledge production.

Between sadness and laughter

Producing knowledge is also always about narrative: about sharing stories, constructing stories, circulating stories. Withholding stories. *Nanay* was constructed to bring audiences closer to the experiences of racialised temporary foreign workers and, in the case of non-Filipino audience members, to build a complicated relationship of witnessing: to feel both empathy and responsibility. Equally, we aimed to bring audiences closer to the mundane, ordinary circumstances of middle-class Canadian families in need of care framed within the broader devaluation of care-work. The travels of *Nanay*, along with the writings of black, Indigenous and postcolonial feminists, have provided resources to think more about the politics of storytelling. It has opened questions about the whiteness of the production: what needed to be learned, what we failed to learn, what needed to be heard and what we could not hear.

Why, Saidiya Hartman asks, does "the site of suffering lend itself to inviting identification?"[25] Beyond the risk of fixing and naturalising this suffering as the condition of being for those who suffer, she, like many others, is suspicious of the benefits that accrue to those who witness another's pain.[26] The repetition and circulation of painful images and suffering can "work to solidify and make continuous the colonial project of violence."[27] In this, empathetic identification can bolster a sense of liberal benevolence, and function as a kind of alibi and as a claim to prior ignorance.[28] Our ambition to educate through performing the testimonials of suffering migrants is likely vulnerable to these critiques, and it is possible that we unwittingly created our own "scenes of subjection."[29]

How to represent racial violence without returning to the wound and profiting from its description? Hartman measures a kind of narrative restraint that refuses to "fill in the gaps and provide closure."[30] This writing does not invite liberal empathy. Quite the contrary. Katherine McKittrick calls for, examines and herself creates modes of writing and creative practice that move outside of and undo the framing and fixing of the suffering black body, typically posed in a binary opposition to the "white western liberated human norm."[31] She interprets NourbeSe Philip's poem, *Zong!*, for instance, as a text that both "iterates anti-black violence within the context of slavery" and produces a "knowledge system that momentarily moves outside itself."[32] Responding to what she terms the "ontological negation" of black people, Christina Sharpe looks to "forms of Black expressive culture [...] that do not seek to explain or resolve the question of this exclusion in terms of assimilation, inclusion or civil or human rights, but rather depict aesthetically the impossibility of such resolutions by representing the paradoxes of blackness within and after the legacies of slavery's denial of Black humanity." She names this paradox "the wake" and the reporting on it the "dysgraphia of disaster."[33] Indigenous scholars are supplementing narratives of victimhood with knowledges of resurgence, resistance and refusal.

We have discussed at length (in Chapter 1) how we attempted, through testimonial monologues, to force audiences into complex, unsettling identifications that might both allow and disrupt empathy. We offer no easy resolution to the question of care. And yet, the Vancouver Children's Choice Awards found *Nanay* the saddest play in the PuSh festival. It appeared that the testimonial monologues solidified the suffering that was already fixed and naturalised on Filipino bodies, such that professional Filipino actors were read by some audience members as being domestic workers in life; in a repetition of the deskilling critiqued in the play, they were stripped of their professional qualifications and disallowed mimetic distance. In a scene with a child of a domestic worker in Canada, it was brought to our attention by the Filipino youths who had interviewed her that we had simplified the emotional tone of the testimony and rendered what she perceived as success to be at least partial failure. We had transformed and typecast her into a simplified subject in pain.

Travelling with the play was a journey in experimenting with ways of sharing the very real suffering that domestic workers' experience under the LCP without allowing this experience to define them, and to disrupt the binary between suffering Filipina and privileged white employers. It also allowed further experimentation on how to represent middle-class Canadian dilemmas about care as more (or other) than the trivial concerns of the privileged, and to explore whether and how to hold them in the same emotional and political space as the predicament of domestic workers. At the PETA Theater in Manila, the director attempted to work away at binaries by introducing video materials from a recorded focus group with domestic workers in Vancouver, who were hanging out and laughing about their employers, and, in another scene, by putting the bodies of employers on display for inspection, that is, by objectifying the employers. There are other possibilities. But what if, as the term racial capitalism signifies, racial violence is the ground that racial capitalism stands on? What does it mean to bring white employers and Filipino domestic workers into the same political space?

Not only the language but the tone changed significantly as we reworked and adapted the play within the histories of Bagong Barrio. From sadness emerged anger, mixed with other emotions: from a young girl sexually assaulted by her grandfather; by a chorus chastising their community for forgetting their recent history of political mobilisation and enjoining them to action; from a group given the opportunity to talk back to the racist monologue of a white, English-speaking nanny agent. There were also expressions of love and hope from a young man telling of his attempts to repair the damages of his childhood through mimed intimate gestures with his mother in a fast-food restaurant in Jeddah. By the time *Nanay* had spawned *Tlingipino Bingo*, it was very clear that there would be no easy resolution of different yet resonant histories of "ontological negation" through gestures of inclusion within national belonging. And there was a new tone: laughter, with different inflections and punch depending on who it was directed towards. Gramma Susie, Miss Lituya and Ardie called the bingo and called the shots. They are not innocent victims calling up sympathy. The days of imagining that our writing, including our academic writing, is toneless, living within an airless room of objectivity, are long gone and there is reason to reflect on the tonality of our writing and how an expanded repertoire can be put to work – in many different creative ways – to disrupt the binary of suffering black body and universal empathetic liberal white subject. The director of *Nanay* in Bagong Barrio and Gramma Susie in Whitehorse offered clear alternatives.

Nanay has never been about inhabiting a shared consciousness through a simplifying absorptive notion of empathy. In the Vancouver production, white audience members reported being most affected by travelling through the show – room to room – alongside Filipino audience members precisely because it brought them to the understanding that they did not inhabit the same lived experience or the same emotional space. It destabilised their privilege of having a view from nowhere. As the play moved and responded to criticism about the representation of Filipino domestic workers, it also became harder for audiences (and for us) to be part of the conversation in an unmarked non-situated way. But, to repeat the question: what if, as the term racial capitalism signifies, racial violence is the ground on which racial capitalism stands? And what does this mean for how white audiences and researchers might be part of the conversation? "[T]he ally is not a stable category," notes Frank Wilderson in reference to a black/white binary in the United States. "There's a structural prohibition (rather than merely a willful refusal) against whites being the allies of blacks [...]: a structural antagonism."[34] Can white allies then play a role in shaking the foundations of racial capitalism and dismantling the binaries through which racial violence works? Beyond marking the violence of racial capitalism, we return to a recurring suggestion that one promising strategy (in which white allies can take responsibility) is to dismantle the coherence of the transparent white subject and displace its "presumed dominion over the lived meanings of race/place/body."[35] Our journey with *Nanay* has been one of documenting and simultaneously expanding and situating perspectives on labour migration, to augment the view from Vancouver within a fuller relational account. We return home with a better understanding of both the place-based particularity of Filipino migration – including within Canada – and a more vital understanding of the need to explore ongoing processes and practices of settler colonialism. This has been made possible by listening carefully – as unstable allies – to criticisms concerning the limits of our perspectives and by inviting close analysis of the ways that we may have unwittingly reproduced the grounds of racial capitalism and colonial relations through our research, artistic practice and the emotional registers in which we work. And by experimenting with ways of doing better.

In the world

In a dense overlap of *Nanay*'s networks, in spring 2018, Hazel Venzon took her new work, *The Places We Go*, to a performance festival in Tacloban City in Leyte, Philippines, at the invitation of the director of *Tlingipino Bingo* (Figure 5.1). Venzon's multimedia performance tells the story of a young girl living in Metro Manila who is attempting to raise enough money for

Figure 5.1 Hazel Venzon performing in *The Places We Go.*

the perfect gift: an air ticket that will allow her mother living and working in Winnipeg, Canada, to return home. The performance is a live graphic novel. Images drawn by co-creator, David Oro, are projected, and as these move across the screen Venzon performs the character of the young girl as a shadow image within the graphic novel.[36] A merging of animate and inanimate, text, image, bodies and voice, a travelling show. *Nanay* has been such a merging and journey that has allowed us to take some steps towards a more global, more local and more intimate understanding of labour migration and to a much-expanded appreciation of the grinding violence of racial capitalism and colonialism in its many iterations. It has also led us to a deep appreciation of the serendipitous openings and possibilities in the world.

Notes

1 David Macdonald, "Child Care Deserts," *Canada Canadian Centre for Policy Alternatives*, 18 June 2018.
2 Organization for Economic Cooperation and Development, "OECD Family Database." Database, Paris, 2017.
3 Organization for Economic Cooperation and Development, and the European Commission, "A Good Life in Old Age? Monitoring and Improving Quality of Long-Term Care," Policy Brief, Paris, June 2013.
4 Pat Armstrong, Barbara Clow, Karen Grant, Margaret Haworth-Brockman, Beth Jackson, Ann Pederson, and Morgan Seeley, eds., *Thinking Women and Health Care Reform in Canada* (Toronto: Women's Press, 2012); Marcy Cohen, "What Matters to Women Working in Long-Term Care: A Union Perspective," in *A Place to Call Home: Long Term Care in Canada*, eds., Pat Armstrong, Madeline Boscoe, Barbara Clow, Karen Grant, Margaret Haworth-Brockman, Beth Jackson, Ann Pederson, Morgan Seeley, and Jane Springer (Winnipeg: Fernwood Publishing, 2009), 97–103; Nicole Molinari, "Bare Care: Working Within/Reworking Restructured Long-Term Care" (unpublished Master's Thesis, University of British

Columbia, 2018); Ann Silversides, "Long-Term Care in Canada: Status Quo No Option" (Canadian Federation of Nurses Unions, Ottawa, 3 February 2011).

5 The narrative extends beyond the LCP. For Canada's self-narrative as a peacekeeper of the world, see Sherene Razack, *Dark Threats and White Knights: The Somalia Affair, Peacekeeping, and the New Imperialism* (Toronto: University of Toronto Press, 2004). As another indicator of Canada's self-perception as compassionate humanitarian, consider Prime Minister Justin Trudeau's words at a rally in Ottawa after his election in 2015: "Many of you have worried that Canada has lost its compassionate and constructive voice in the world over the past 10 years. Well, I have a simple message for you: on behalf of 35 million Canadians, we're back" (quoted in *Canadian Press*, "'We're back,' Justin Trudeau says in message to Canada's allies abroad," *National Post*, 20 October 2015). For a popular presentation of the stereotypes that Canadian hold of themselves relative to Americans (including the superiority of Canadian humanitarianism and healthcare system), see Lilly Singh's "What Canadians Really Want to Say to Americans": www.youtube.com/watch?v=bcmbKN1aIOU.

6 If anything, the issues raised by the play have only intensified. See note 7 of Introduction.

7 Robyn Magalit Rodriguez, "Toward a Critical Filipino Approach to Philippine Migration," in *Filipino Studies: Palimpsests of Nation and Diaspora*, eds. Martin F. Manalanson IV and Augusto F. Espiritu (New York: New York University Press, 2016), 42.

8 This is not to romanticise the Philippines, especially in the current climate of vigilante violence, encouraged and supported by President Duterte: for example, Human Rights Watch World Report 2018, "Philippines Events of 2017," www.hrw.org/world-report/2018/country-chapters/philippines.

9 Witness the World Bank's appreciative note that flows of remittances to the Philippines "remained buoyant" in 2016 as compared to the trend, defined as risk, for developing countries more generally (World Bank Group, "Migration and Remittances: Recent Developments and Outlook," World Bank Migration and Development Brief 27, Washington DC, 2017).

10 For a discussion of the refusal of the posthumous award of U.S. citizenship by spouses of Filipino men who have died fighting in the U.S. military, see Elizabeth Lee, "Foreign Born Soldiers and the Ambivalent Spaces of Citizenship" (unpublished PhD dissertation, University of British Columbia, 2012).

11 May Farrales, "Gendered Sexualities in Migration: Play, Pageantry, and the Politics of Performing Filipino-ness in Settler Colonial Canada" (unpublished PhD dissertation, University of British Columbia, 2017).

12 Alongside spreading anti-immigrant rhetoric expressed by groups more readily labelled as racist, white supremacist or populist, consider recent debates around two Oxford academics' suggestion of a place called "Refugia" – a self-governing autonomous settlement archipelago in which refugees would live and work – as a solution to the refugee crisis, along with similar ideas for "Zatopia," "Refugee Nation," "Europe-in-Africa," etc. (Robin Cohen, "Refugia: A Utopian Solution to the Crisis of Mass Displacement," *Conversation*, 7 August 2017; for a critique of this and similar notions, see Helen Crawley, "Why We Need to Protect Refugees from the 'Big Ideas' Designed to Save them," *Independent*, 28 July 2018). We thank Jennifer Hyndman for drawing this strand of thinking to our attention.

13 See note 7 of Introduction.

14 Migrante BC (along with other organisations, such as the Migrant Workers Centre in Vancouver and No One Is Illegal) has been concerned about the increased policing of domestic workers alongside other TFWs. Project Guardian, for instance, was launched as a pilot project by the CBSA in British Columbia and the

Yukon in January 2014 in the wake of reports that the LCP was being abused by domestic workers who were violating the terms of their contract. Domestic workers are now much more cautious about working outside their contract, at extra jobs or while waiting for new contracts. The numbers of immigration investigations initiated by the CBSA Pacific region from 2013 through 2015 were **2413** in 2013, **2563** in 2014 and **2425** in 2015 (Travis Lupick, "CBSA Enforcement Actions in B.C. Hold Steady Despite Sanctuary City Gains Across Metro Vancouver," *The Georgia Straight*, 22 February 2017). As of October 2015, 38 caregivers had been investigated under the Project Guardian initiative (see West Coast Domestic Workers' Association, "Temporary Foreign Worker Program: A Submission by the West Coast Domestic Workers' Association to the Standing Committee on Human Resources, Skills and Social Development and the Status of Persons with Disabilities," 31 May 2016).

15 Gilberto Rosas, *Barrio Libre: Criminalizing States and Delinquent Refusals of the New Frontier* (Durham and London: Duke University Press, 2012), 109.

16 Audra Simpson, "Consent's Revenge," *Cultural Anthropology* 31, no. 3 (2016): 331.

17 Carol McGrahahan, "Refusal and the Gift of Citizenship," *Cultural Anthropology* 31, no. 3 (2016): 334–341.

18 For some of the varied responses and strategies deployed by Indigenous activists in Australia and New Zealand in response to recent arrivals of refugees, asylum seekers and overstayers, see Erica Weiss, "Refusal as Act, Refusal as Abstention," *Cultural Anthropology* 31, no. 3 (2016): 351–358; Emma Cox, "Sovereign Ontologies in Australia and Aotearoa-New Zealand: Indigenous Responses to Asylum Seekers, Refugees and Overstayers," in *Knowing Differently: The Challenges of the Indigenous*, eds. G.N. Devy, Geoffrey Davis, and K.K. Chakravarty (New Delhi: Routledge, 2013), 139–157.

19 Emilie Cameron, *Far Off Metal River: Inuit Lands, Settler Stories, and the Making of the Contemporary Arctic* (Vancouver: University of British Columbia Press, 2015), 17.

20 If you would like to join in the fun, see http://joledingham.ca/penelope.

21 Richa Nagar, "Hungry translations: the world through radical vulnerability," *Antipode* early view (2018): 16.

22 Emilie Cameron, *Far Off Metal River: Inuit Lands, Settler Stories, and the Making of the Contemporary Arctic* (Vancouver: University of British Columbia Press, 2015).

23 As Sarah de Leeuw and Sarah Hunt note, the concept of decolonisation is "complicated and contested" (Sarah de Leeuw and Sarah Hunt, "Unsettling Decolonizing Geographies," *Geography Compass* 12, no. 7 (2018): 2) and there is the ever-present danger of it becoming another means of re-centring settler voices and perspectives as the concept is "emptied of its substance and instrumentalized by settler researchers and institutions" (Hugo Asselin and Suzy Basile, "Concrete Ways to Decolonize Research," *ACME: An International Journal for Critical Geographies* 17, no. 3 (2018): 645; see also Sarah de Leeuw, Emilie Cameron, and Margo Greenwood, "Participatory and Community-Based Research, Indigenous Geographies, and the Spaces of Friendship: A Critical Engagement," *The Canadian Geographer* 56, no. 2 (2012): 180–194; Eve Tuck and K. Wayne Yang, "Decolonization Is Not a Metaphor," *Decolonization: Indigeneity, Education & Society* 1, no. 1 (2012): 1–40). Ideally, decolonial practices involve moving away from a long history of non-Indigenous scholars attempting to define Indigenous reality, towards mutual recognition, reciprocity, accountability to Indigenous sovereignties and futurity, and to modes of enquiry that attend to process as much as outcome. Arguing for the need to unsettle decolonising geographies, de Leeuw and Hunt call for non-Indigenous scholars to unsettle our own authority

by attending to citational practices, by politicising our own situated position on stolen or colonised land, and reflecting on the pervasive impacts of, along with personal benefits from, the ongoing colonial violences in the places that we study and in which we work and live.

24 Shu-mei Shih, "World Studies and Relational Comparison," *PMLA* 130, no. 2 (2015): 430–438.

25 Saidiya Hartman, *Scenes of Subjection: Terror Slavery, and Self-Making in Nineteenth-Century America* (Oxford: Oxford University Press, 1997), 20.

26 Writing about histories of slavery in the United States, Hartman argues that the exposure of the black suffering body reinforced the "'thingly' quality of the captive body" (Ibid., 19) and invited empathetic voyeurism that was entangled with possession.

27 Christina Sharpe, *In the Wake: On Blackness and Being* (Durham and London: Duke University Press, 2016), 117.

28 "If only they knew the truth, they would act otherwise" (Saidiya Hartman, *Lose Your Mother: A Journey along the Atlantic Slave Route* (New York: Farrar, Straus and Giroux, 2007), 169). Hartman is reminded of James Baldwin, writing to his nephew on the centenary of the Emancipation Proclamation: "It is not permissible that the authors of devastation should also be innocent. It is the innocence which constitutes the crime" (quoted in Ibid., 169).

29 Saidiya Hartman, *Scenes of Subjection: Terror Slavery, and Self-Making in Nineteenth-Century America* (Oxford: Oxford University Press, 1997).

30 Saidiya Hartman, "Venus in Two Acts," *Small Axe* 12, no. 2 (2008): 12.

31 Katherine McKittrick, "Diachronic Loops/Deadweight Tonnage/Bad Made Measure," *Cultural Geographies* 23, no. 1 (2016): 16.

32 Ibid., 13–14.

33 Christina Sharpe, *In the Wake: On Blackness and Being* (Durham and London: Duke University Press, 2016), 20, 21; these discussions resonate with Audra Simpson's ethnographic refusal. Writing as an Indigenous anthropologist about the people of Kahnawàke, Simpson has refused to be "*that* thick description prosemaster who would reveal in florid detail" the internal story of their struggle. In this stance of ethnographic refusal, she will only tell "the story of their constraint" (Audra Simpson, "Consent's Revenge," *Cultural Anthropology* 31, no. 3 (2016): 328).

34 Saidiya Hartman and Frank B. Wilderson III, "The Position of the Unthought," *Qui Parle* 13, no. 2 (2007): 190.

35 Dylan Rodriguez, *Suspended Apocalypse: White Supremacy, Genocide, and the Filipino Condition* (Minneapolis: University of Minnesota Press, 2010), 199.

36 For more detail on *The Places We Go*, see http://unitprod.ca/the-places-we-go/.

Bibliography

Abada, Teresa, Feng Hou, and Bali Ram. "Ethnic Differences in Educational Attainment among the Children of Canadian Immigrants." *Canadian Journal of Sociology* 34, no. 1 (2009): 1–29.

Abraham, Nicholas, and Maria Torok. *The Shell and the Kernel: Renewals of Psychoanalysis*, edited and translated by Nicholas T. Rand. Chicago: University of Chicago Press, 1994.

Advincula-Lopez, Leslie V. "OFW Remittances, Community, Social and Personal Services and the Growth of Social Capital." *Philippine Sociological Review* 53 (2005): 58–74.

Aguilar, Filomeno. "Brother's Keeper? Siblingship, Overseas Migration, and Centripetal Ethnography in a Philippine Village." *Ethnography* 14, no. 3 (2013): 346–368.

Aguilar, Filomeno. *Migration Revolution: Philippine Nationhood and Class Relations in a Globalized Age.* Singapore and Kyoto: National University of Singapore Press in association with Kyoto University Press, 2014.

Ahmed, Sara. "A Phenomenology of Whiteness." *Feminist Theory* 8, no. 2 (2007): 149–168.

Alarcon, Krystle. "First Filipina Nanny Helped Community Grow." *Yukon News*, 30 August 2013. www.yukon-news.com/life/first-filipina-nanny-helped-community-grow/.

Alcoff, Linda. "The Problem of Speaking for Others." *Cultural Critique* 20 (1992): 5–32.

Anderson, Kay, and Susan J. Smith. "Editorial: Emotional Geographies." *Transactions of the Institute of British Geographers* 26, no. 1 (2001): 7–10.

Anderson, Michael, and Linden Wilkinson. "A Resurgence of Verbatim Theatre: Authenticity, Empathy, and Transformation." *Australasian Drama Studies* 50 (2007): 153–169.

Angeles, Leonora, and Geraldine Pratt. "Empathy and Entangled Engagements: Critical-Creative Methodologies in Transnational Spaces." *GeoHumanities* 3, no. 2 (2017): 269–278.

Antwi, Phanuel, Sarah Brophy, Helene Strauss, and Y-Dang Troeung. "'Not Without Ambivalence': An Interview with Sara Ahmed on Postcolonial Intimacies." *Interventions: International Journal of Postcolonial Studies* 15, no. 1 (2013): 110–126.

Arat-Koç, Sedef. "An Anti-Colonial Politics of Place." In "Intervention – Addressing the Indigenous-Immigration 'Parallax Gap,'" edited by Anna Stanley, Sedef Arat-Koc, Laurie Bertram, and Hayden King, n.p. *Antipode Foundation* (blog), 18 June 2014. https://antipodefoundation.org/2014/06/18/addressing-the-indigenous-immigration-parallax-gap/.

Arat-Koç, Sedef. "Politics of the Family and Politics of Immigration in the Subordination of Domestic Workers in Canada." In *Family Patterns, Gender Relations*, 2nd edition, edited by Bonnie Fox, 352–374. Toronto: Oxford University Press, 1993.

Arendt, Hannah. "Introduction: Walter Benjamin: 1892–1940." In *Illuminations*, edited by Walter Benjamin, 1–55. Translated by Harry Zohn. New York: Schocken Books, 1969.

Armstrong, Pat, Barbara Clow, Karen Grant, Margaret Haworth-Brockman, Beth Jackson, Ann Pederson, and Morgan Seeley, eds. *Thinking Women and Health Care Reform in Canada*. Toronto: Women's Press, 2012.

Asis, Maruja. "The Philippines: Beyond Labor Migration, Toward Development and (Possibly) Return." *Migration Policy Institute* (blog), 12 June 2017. www.migrationpolicy.org/article/philippines-beyond-labor-migration-toward-development-and-possibly-return.

Asselin, Hugo, and Suzy Basile. "Concrete Ways to Decolonize Research." *ACME: An International Journal for Critical Geographies* 17, no. 3 (2018): 643–650.

Bagasao, Ildefonso F., Elena B Piccio, Lourdes T. Lopez, and Peter Djinis. "Enhancing the Efficiency of Overseas Filipino Workers Remittances." Asian Development Bank, Manila, 2004. www.microfinancegateway.org/sites/default/files/mfg-en-paper-enhancing-the-efficiency-of-overseas-workers-remittances-2004.pdf.

Bakan, Abigail, and Daiva Stasiulis. *Not One of the Family: Domestic Workers in Canada*. Toronto: University of Toronto Press, 1997.

Banta, Vanessa. "Empathetic Projections: Performance and Countermapping of *Sitio* San Roque, Quezon City, and University of the Philippines." *GeoHumanities* 3, no. 2 (2017): 328–350.

Baucom, Ian. *Specters of the Atlantic: Finance Capital, Slavery, and the Philosophy of History*. Durham: Duke University Press, 2005.

Belliveau, George, and Graham Lea, eds. *Research-Based Theatre: An Artistic Methodology*. Bristol: Intellect, 2016.

Bello, Walden, Marissa de Guzman, Mary L. Malig and Herbert Docena. *The Anti-Development State: The Political Economy of Permanent Crisis in the Philippines*. London: Zed Books, 2005.

Benedicto, Bobby. *Under Bright Lights: Gay Manila and the Global Scene*. Minnesota: University of Minnesota Press, 2014.

Ben-Zvi, Linda. "Staging the Other Israel: The Documentary Theatre of Nola Chilton." *The Drama Review* 50, no. 3 (2006): 42–55.

Berlant, Lauren. "Critical Inquiry, Affirmative Culture." *Critical Inquiry* 30, no. 2 (2004): 445–451.

Berlant, Lauren. *Cruel Optimism*. Durham: Duke University Press, 2011.

Berlant, Lauren. "Thinking about Feeling Historical." *Emotion, Space and Society* 1, no. 1 (2008): 4–9.

Bernstein, Robin. *Racial Innocence: Performing Childhood from Slavery to Civil Rights*. New York: New York University Press, 2011.

Bertram, Laurie. "Icelandic and Indigenous Exchange, Overlapping Histories of Migration and Colonialism." In *Intervention – Addressing the Indigenous-Immigration 'Parallax Gap'*, edited by Anna Stanley, Sedef Arat-Koc, Laurie Bertram, and Hayden King, n.p. *Antipode Foundation* (blog), 18 June 2014. https://antipodefoundation.org/2014/06/18/addressing-the-indigenous-immigration-parallax-gap/.

Bertram, Laurie. "Resurfacing Landscapes of Trauma: Multiculturalism, Cemeteries, and the Migrant Body, 1875 Onwards." In *Home and Native Land: Unsettling Multiculturalism in Canada*, edited by May Chazon, Lisa Helps, Anna Stanley, and Sonali Thakkar, 157–174. Toronto: Between The Lines, 2011.

Beyes, Timon. "Uncontained: The Art and Politics of Reconfiguring Urban Space." *Culture and Organization* 16, no. 3 (2010): 229–246.

Bishop, Clare. *Artificial Hells: Participatory Art and the Politics of Spectatorship*. London: Verso, 2012.

Blunt, Alison, Jayani Bonnerjee, Caron Lipman, Joanne Long, and Felicity Paynter. "My Home: Text, Space, and Performance." *Cultural Geographies* 14 (2007): 309–318.

Boltanski, Luc. *Distant Suffering: Mortality, Media, and Politics*. Cambridge: Cambridge University Press, 1999.

Bonus, Rick. "'Come Back Home Soon': The Pleasures and Agonies of 'Homeland' Visits." In *Filipino Studies: Palimpsests of Nation and Diaspora*, edited by Martin F. Manalanson IV and Augusto F. Espiritu, 388–410. New York: New York University Press, 2016.

Boyer, Dominic, and Cymene Howe. "Portable Analytics and Lateral Theory." In *Theory Can Be More Than It Used to Be*, edited by Dominic Boyer, James D. Faubion and George E. Marcus, 15–38. Ithaca, NY: Cornell University Press, 2015.

Brydon, Anne. "Dreams and Claims: Icelandic-Aboriginal Interactions in the Manitoba Interlake." *Journal of Canadian Studies* 36, no. 2 (2001): 164–190.

Burk, Adrienne L. *Speaking for a Long Time: Public Space and Social Memory in Vancouver*. Vancouver: University of British Columbia Press, 2010.

Burns, Lucy Mae San Pablo. *Puro Arte: Filipinos on the Stages of Empire*. New York and London: New York University Press, 2012.

Burvill, Tom. "'Politics Begins as Ethics': Levinasian Ethics and Australian Performance Concerning Refugees." *Research in Drama Education* 13, no. 2 (2008): 233–243.

Butler, Judith. *Giving an Account of Oneself*. New York: Fordham University Press, 2005.

Butler, Judith. *Notes towards a Performative Theory of Assembly*. Cambridge, MA: Harvard University Press, 2015.

Byrd, Jodi. *Transit of Empire: Indigenous Critiques of Colonialism*. Minneapolis and London: University of Minnesota Press, 2011.

Çaglar, Ayse, and Nina Glick Schiller. *Migrants and City-Making: Dispossession, Displacement, and Urban Regeneration*. Durham and London: Duke University Press, 2018.

Callon, Michel, and Vololona Rabeharisoa. "Research 'in the Wild' and the Shaping of New Social Identities." *Technology and Society* 25, no. 2 (2003): 193–204.

Cameron, Emilie. *Far Off Metal River: Inuit Lands, Settler Stories, and the Making of the Contemporary Arctic*. Vancouver: University of British Columbia Press, 2015.

Campomanes, Oscar V. "New Formations in Asian American Studies and the Question of U.S. Imperialism." *positions: asia critique* 5, no. 2 (1997): 523–550.

Canadian Press. "'We're Back,' Justin Trudeau Says in Message to Canada's Allies Abroad." *National Post*, 20 October 2015. https://nationalpost.com/news/politics/were-back-justin-trudeau-says-in-message-to-canadas-allies-abroad.

Cannell, Fenella. *Power and Intimacy in the Christian Philippines*. Cambridge: Cambridge University Press, 1999.

Caraballo, Mayvelin. "OFW Remittances Hit New High." *Manila Times*, 16 February 2015. www.manilatimes.net/ofw-remittances-hit-new-high/163522.

Chakarbarty, Dipesh. *Provincializing Europe: Postcolonial Thought and Historical Difference*. Princeton: Princeton University Press, 2000.

Chazon, May, Lisa Helps, Anna Stanley, and Sonali Thakkar, eds. *Home and Native Land: Unsettling Multiculturalism in Canada*. Toronto: Between The Lines, 2011.

Citizenship and Immigration Canada. "Immigrant Overview – Permanent Residents." Canada Facts and Figures, Ottawa, 2014. www.cic.gc.ca/english/pdf/2014-Facts-Permanent.pdf.

Clarkson, Joshua. "Foreign Workers Take Jobs that Nobody else Wants." *Yukon News*, 11 March 2015. www.yukon-news.com/news/foreign-workers-take-jobs-that-nobody-else-wants/.

Cohen, Marcy. "What Matters to Women Working in Long-Term Care: A Union Perspective." In *A Place to Call Home: Long Term Care in Canada*, edited by Pat Armstrong, Madeline Boscoe, Barbara Clow, Karen Grant, Margaret Haworth-Brockman, Beth Jackson, Ann Pederson, Morgan Seeley, and Jane Springer, 97–103. Winnipeg: Fernwood Publishing, 2009.

Cohen, Robin. "Refugia: A Utopian Solution to the Crisis of Mass Displacement" *The Conversation*, 7 August 2017. https://theconversation.com/refugia-a-utopian-solution-to-the-crisis-of-mass-displacement-81136.

Collins, Erin. "Of Crowded Histories and Urban Theory" Unpublished paper.

Coloma, Roland Sintos, Bonnie McElhinny, Ethel Tungohan, John Paul Catungal, and Lisa M. Davidson, eds. *Filipinos in Canada: Disturbing Invisibility*. Toronto: University of Toronto Press, 2012.

Conquergood, Dwight. "Rethinking Ethnography: Towards a Critical Cultural Politics." In *The SAGE Handbook of Performance Studies*, edited by D. Soyini Madison and Judith Hamera, 351–365. London: Sage, 2006.

Constantino, Renato. "The Miseducation of the Filipino." In *Vestiges of War: The Philippine–American War and the Aftermath of an Imperial Dream 1899–1999*, edited by Angel Velasco Shaw and Luis H. Francia, 177–192. New York: New York University Press, 2002.

Cox, Emma. "Sovereign Ontologies in Australia and Aotearoa-New Zealand: Indigenous Responses to Asylum Seekers, Refugees and Overstayers." In *Knowing Differently: The Challenges of the Indigenous*, edited by G.N. Devy, Geoffrey Davis, and K.K. Chakravarty, 139–157. New Delhi: Routledge, 2013.

Cox, Emma. *Staging Asylum: Contemporary Australian Plays about Refugees*. Redfern: Currency Press, 2013.

Cox, Emma. *Theatre & Migration*. New York: Palgrave Macmillan, 2014.

Cox, Emma, and Caroline Wake. "Envisioning Asylum/Engendering Crisis: Or, Performance and Forced Migration 10 years on." *Research in Drama Education* 23, no. 2 (2018): 137–147.

Crawley, Helen. "Why We Need to Protect Refugees from the 'Big Ideas' Designed to Save Them." *Independent*, 28 July 2018. www.independent.co.uk/voices/refugee-immigration-europe-migrants-refugia-self-governance-a8467891.html.

Cresswell, Tim. "Displacements – Three Poems." *Geographical Review* 103, no. 2 (2013): 285–287.

Crutchlow, Paula, and Helen V. Jamieson. "Make-Shift." *Liminalities: A Journal of Performance Studies* 10, no. 1 (2014): n.p.

Curry, Bill. "Live-In Caregivers May Be Next Target of Immigration Reform." *Globe and Mail*, 23 June 2014. www.theglobeandmail.com/news/politics/live-in-caregivers-may-be-next-target-of-immigration-reform/article19304029/.

Da Costa, Dia. "Eating Heritage: Caste, Colonialism and the Contestation of Indigenous Creativity." Paper presented at Reimagining Creative Economies workshop, Edmonton, Alberta, April 2017.

Da Costa, Dia. *Politicizing Creative Economy: Activism and a Hunger Called Theater*. Urbana: University of Illinois, 2016.

Daenzer, Patricia. *Regulating Class Privilege: Immigrant Servants in Canada, 1940s–1990s*. Toronto: Canadian Scholars' Press, 1993.

Davidson, Lisa. "(Res)sentiment and Practices of Hope: The Labours of Filipina Live-In Caregivers in Filipino Canadian Families." In *Filipinos in Canada: Disturbing Invisibility*, edited by Roland Sintos Coloma, Bonnie McElhinny, Ethel Tungohan, John Paul Catungal, and Lisa M. Davidson, 142–160. Toronto: University of Toronto Press, 2012.

Day, Iyko. *Alien Capital: Asian Racialization and the Logic of Settler Colonial Capitalism*. Durham and London: Duke University Press, 2016.

Dean, Jodi. "Politics without Politics." *Parallax* 15, no. 3 (2009): 20–36.

De Lauretis, Teresa. *Technologies of Gender: Essays on Theory, Film, and Fiction*. Bloomington: Indiana University Press, 1987.

De Leeuw, Sarah. *Geographies of a Lover*. Edmonton: NeWest Press, 2012.

De Leeuw, Sarah. *Where It Hurts*. Edmonton: NeWest Press, 2017.

De Leeuw, Sarah, Emilie Cameron, and Margo Greenwood. "Participatory and Community-Based Research, Indigenous Geographies, and the Spaces of Friendship: A Critical Engagement." *The Canadian Geographer* 56, no. 2 (2012): 180–194.

De Leeuw, Sarah, and Harriet Hawkins. "Critical Geographies and Geography's Creative Turn: Poetics and Practices for New Disciplinary Spaces." *Gender, Place and Culture: A Journal of Feminist Geography* 24, no. 3 (2017): 303–324.

De Leeuw, Sarah, and Sarah Hunt. "Unsettling Decolonizing Geographies." *Geography Compass* 12, no. 7 (2018): 1–14.

De Souza, Raymond. "The Yukon is the North, Pure and Simple." *National Post*, 7 June 2012. https://nationalpost.com/opinion/father-raymond-j-de-souza-the-yukon-is-the-north-pure-and-simple.

Deloria, Philip J. *Indians in Unexpected Places*. Lawrence: University of Kansas Press, 2004.

Dikec, Mustafa. "Beginners and Equals: Political Subjectivity in Arendt and Rancière." *Transactions of the Institute of British Geographers* 38, no. 1 (2013): 78–90.

Dillon, Matthew. *Pilgrims and Pilgrimage in Ancient Greece*. New York: Routledge, 1997.

Dolan, Jill. *Utopia in Performance: Finding Hope at the Theater*. Ann Arbor: University of Michigan Press, 2005.

Dreby, Joanna. *Divided by Borders: Mexican Migrants and Their Children*. Berkeley: University of California Press, 2010.

England, Kim, and Bernadette Stiell. "'They Think You're as Stupid as Your English Is': Constructing Foreign Domestic Workers in Toronto." *Environment and Planning A* 29, no. 2 (1997): 195–215.

Everett-Green, Robert. "Kent Monkman: A Trickster with a Cause Crashes Canada's 150th Birthday Party." *Globe and Mail*, 7 January 2017. www.theglobeandmail.com/news/national/canada-150/kent-monkman-shame-and-prejudice/article33515775/.

Fagan, Kristina. "Laughing to Survive: Humour in Contemporary Canadian Native Literature." Unpublished PhD Dissertation, University of Toronto, 2001.

Farrales, May. "Delayed, Deferred and Dropped Out: Geographies of Filipino-Canadian High School Students." *Children's Geographies* 15, no. 2 (2017): 207–223.

Farrales, May. "Gendered Sexualities in Migration: Play, Pageantry, and the Politics of Performing Filipino-ness in Settler Colonial Canada." Unpublished PhD Dissertation, University of British Columbia, 2017.

Ferguson, Alex Lazaridis. "Authenticity and the 'Documentative' in Nanay: A Testimonial Play." *Platform: Journal of Theatre and Performing Arts* 11 (2017): 88–109.

Ferguson, Alex Lazaridis. "Improvising the Document." *Canadian Theatre Review* 143 (2010): 35–41.

Fernandez, Doreen. *Palabas: Essays on Philippine Theatre History.* Manila: Ateneo University Press, 1996.

Fischer-Lichte, Erika. *The Transformative Power of Performance: A New Aesthetics.* New York: Routledge, 2008.

Fisek, Emine. *Aesthetic Citizenship: Immigration and Theater in Twenty-First-Century Paris.* Evanston: Northwestern University Press, 2017.

Frontline. *Private Warriors.* Directed by Tim Mangini. 2005; Boston: Public Broadcast Service. Video Broadcast. www.pbs.org/wgbh/pages/frontline/shows/warriors.

Gabie, Neville, Joan Gabie, and Ian Cook. "Dust, in 'Bideford Black, the Next Generation.'" Exhibition at the Burton Art Gallery & Museum, Bideford, 3 October–13 November2015.www.nevillegabie.com/exhibitions-archive/bideford-black-the-next-generation/.

Gallagher, Michael. "Sounding Ruins: Reflections on the Production of an 'Audio Drift'." *Cultural Geographies* 22, no. 3 (2015): 467–485.

Garrido, Marco. "The Ideology of the Dual City: The Modernist Ethic in the Corporate Development of Makati City, Metro Manila." *International Journal of Urban and Regional Research* 37, no. 1 (2013): 165–185.

Gidwani, Vinay, and Rajyashree Reddy. "The Afterlives of "Waste": Notes from India for a Minor History of Capitalist Surplus." *Antipode* 43, no. 5 (2011): 1625–1658.

Gilbert, Helen, and Jacqueline Lo. *Performance and Cosmopolitics: Cross-Cultural Transactions in Australasia.* Basingstoke: Palgrave Macmillan, 2007.

Global Commission on International Migration. "Global Migration for an Interconnected World: New Directions for Action." Report, Switzerland, 2005.

Gonzalez, Andrew. *Language and Nationalism: The Philippine Experience Thus Far.* Quezon City: Ateneo de Manila University Press, 1980.

Government of Canada. "Temporary Foreign Worker Program Labour Market Impact Assessment (LMIA) Statistics Fourth Quarter, 2015." Quarterly Labour Market Impact Assessment Statistics, Ottawa, 2016. www.canada.ca/en/employment-social-development/services/foreign-workers/2015-quarterly-labour-market-information.html.

Guevarra, Anna R. *Marketing Dreams, Manufacturing Heroes: The Transnational Labor Brokering of Filipino Workers.* New Brunswick, NJ: Rutgers University Press, 2010.

Guevarra, Anna R. "Supermaids: The Racial Branding of Global Filipino Care Labour." In *Migration and Care Labour: Theory, Policy and Politics*, edited by Bridget Anderson and Isabel Shutes, 130–150. Houndmills: Palgrave Macmillan, 2014.

Guillermo, Alice. *Protest/Revolutionary Art in the Philippines, 1970–1990*. Quezon City: University of the Philippines Press, 2001.

Hallward, Peter. "Staging Equality: On Rancière's Theatrocracy." *New Left Review* 37 (2006): 109–129.

Hammond, Will, and Dan Steward, eds. *Verbatim Verbatim: Contemporary Documentary Theatre*. London: Oberon, 2008.

Hart, Gillian. "Relational Comparison Revisited: Marxist Postcolonial Geographies in Practice." *Progress in Human Geography* 42, no. 3 (2018): 371–394.

Hartman, Saidiya. *Lose Your Mother: A Journey Along the Atlantic Slave Route*. New York: Farrar, Straus and Giroux, 2007.

Hartman, Saidiya. *Scenes of Subjection: Terror, Slavery, and Self-Making in Nineteenth Century America*. Oxford: Oxford University Press, 1997.

Hartman, Saidiya. "Venus in Two Acts." *Small Axe* 26 (2008): 1–14.

Hartman, Saidiya, and Frank B. Wilderson III. "The Position of the Unthought." *Qui Parle* 13, no. 2 (2007): 183–201.

Hau, Caroline Sy. "Privileging Roots and Routes: Filipino Intellectuals and the Contest over Epistemic Power and Authority." *Philippine Studies: Historical and Ethnographic Viewpoints* 62, no. 1 (2014): 29–65.

Hawkins, Harriet. *For Creative Geographies: Geography, Visual Arts and the Making of Worlds*. New York: Routledge, 2014.

Hawkins, Harriet, and Elizabeth Straughan, eds. *Geographical Aesthetics: Imagining Space, Staging Encounters*. London: Routledge, 2015.

Hazou, Rand. "Performing Manaaki and New Zealand Refugee Theatre." *Research in Drama Education* 23, no. 2 (2018): 228–241.

Hedman, Eva-Lotta E., and John T. Sidel. *Philippine Politics and Society in the Twentieth Century: Colonial Legacies, Post-Colonial Trajectories*. London: Routledge, 2000.

Highway, Tomson. *The Rez Sisters: A Play in Two Acts*. Markham: Fifth House, 1990.

Hill, Richard. "Drag Racing: Dressing Up (and Messing Up) White in Contemporary First Nations Art." *Fuse Magazine* 23, no. 4 (2001): 18–27.

hooks, bell. *Yearning: Race, Gender, and Cultural Politics*. Boston: South End Press, 1990.

Honig, Bonnie. "Immigrant America? How Foreignness 'Solves' Democracy's Problems." *Social Text* 56 (1998): 1–27.

Hopper, Tristin. "Foreign Workers Championed Despite Slowdown." *Yukon News*, 3 April 2009. www.yukon-news.com/news/foreign-workers-championed-despite-slowdown/.

Hough, Jennifer. "Canada's Live-In Caregiver Program 'Ran Out of Control' and Will be Reformed: Jason Kenney." *National Post*, 24 June 2014. https://nationalpost.com/news/politics/canadas-live-in-caregiver-program-ran-out-of-control-and-will-be-reformed-jason-kenney.

Hunt, Sarah, and Cindy Holmes. "Everyday Decolonization: Living a Decolonizing Queer Politics." *Journal of Lesbian Studies* 19 (2015): 154–172.

Hutchcroft, Paul. *Booty Capitalism: The Politics of Banking in the Philippines.* Ithaca, NY: Cornell University Press, 1998.

Ingram, Alan. "Rethinking Art and Geopolitics Through Aesthetics: Artist Responses to the Iraq War." *Transactions of the Institute of British Geographers* 41, no. 1 (2016): 1–13.

International Trade Union Confederation. "Facilitating Exploitation: A Review of Labour Laws for Migrant Domestic Workers in Gulf Cooperation Council Countries." Legal and Policy Brief, Brussels, 2014. www.ituc-csi.org/IMG/pdf/gcc_legal_and_policy_brief_domestic_workers_final_text_clean_282_29.pdf.

Jackson, Shannon. *Social Works: Performing Art, Supporting Publics.* New York and London: Routledge, 2011.

Jestrovic, Silvija. "Performing Like an Asylum Seeker: Paradoxes of Hyper-Authenticity." *Research in Drama Education* 13, no. 4 (2008): 159–170.

Johansson, Ola. "The Limits of Community-Based Theatre: Performance and HIV Prevention in Tanzania." *The Drama Review* 54, no. 1 (2010): 59–75.

Johnston, Caleb, and Geraldine Pratt, in Collaboration with the Philippine Women Centre of British Columbia. "Nanay [Mother]: A Testimonial Play." *Cultural Geographies* 17, no. 1 (2010): 123–133.

Johnston, Caleb, and Geraldine Pratt. "Taking Nanay to the Philippines: Transnational Circuits of Affect." In *Theatres of Affect*, edited by Erin Hurley, 192–212. Toronto: University of Toronto Press, 2014.

Johnston, Denis. "Lines and Circles: The 'Rez' Plays of Tomson Highway." Canadian Literature 124-25 (1990): 254–64.

Keevil, Genesee. "The Philippine Connection: How One Woman Turned Whitehorse into a Promised Land." *Walrus*, 18 May 2016. https://thewalrus.ca/the-philippine-connection/.

Kelly, Philip. "Understanding Intergenerational Social Mobility: Filipino Youth in Canada." *IRPP Study* 45 (2014): n.p. http://irpp.org/wp-content/uploads/assets/research/diversity-immigration-and-integration/filipino-youth/kelly-feb-2014.pdf.

Kelly, Philip, Stella Park, Conely de Leon, and Jeff Priest. "Profile of Live-In Caregiver Immigrants to Canada, 1993–2009." TIEDI (Toronto Immigrant Employment Data Initiative) Analytical Report 18, Toronto, 2011. www.yorku.ca/tiedi/doc/AnalyticalReport18.pdf.

Kester, Grant. *The One and the Many: Contemporary Collaborative Art in a Global Context.* Durham and London: Duke University Press, 2011.

Keung, Nicholas. "Foreign Caregivers Face Lengthy Wait for Permanent Resident Status." *Toronto Star*, 21 July 2015. www.thestar.com/news/immigration/2015/07/21/foreign-caregivers-face-lengthy-wait-for-permanent-status.html.

Knowles, Ric. *Theatre and Interculturalism.* New York: Palgrave Macmillan, 2010.

Koo, Jah-Hon, and Jill Hanley. "Migrant Live-In Caregivers: Control, Consensus, and Resistance in the Workplace and the Community." In *Unfree Labour? Struggles of Migrant and Immigrant Workers in Canada*, edited by Aziz Choudry and Adrian Smith, 37–53. Oakland: PM Press, 2016.

Kondo, Dorinne. "Bad Girls: Theatre, Women of Colour and the Politics of Representation." In *Women Writing Culture*, edited by Ruth Behar and Deborah Gordon, 49–64. Berkeley: University of California Press, 1995.

Kondo, Dorinne. "The Narrative Production of 'Home', Community, and Political Identity in Asian American Theatre." In *Displacement, Diaspora and Geographies of Identity*, edited by Smadar Lavie and Ted Swendenburg, 97–117. Durham: Duke University Press, 1996.

Kondo, Dorinne. "(Re)visions of Race: Contemporary Race Theory and the Cultural Politics of Racial Crossover in Documentary Theatre." *Theatre Journal* 52 (2000): 81–107.

Kuokkanen, Rauna. *Reshaping the University: Responsibility, Indigenous Epistemes, and the Logic of the Gift*. Vancouver: University of British Columbia Press, 2007.

Lawrence, Bonita, and Enakshi Dua. "Decolonizing Antiracism." *Social Justice* 32, no. 4 (2005): 120–143.

Lee, Elizabeth. "Foreign Born Soldiers and the Ambivalent Spaces of Citizenship." Unpublished PhD Dissertation, University of British Columbia, 2012.

Lee, Elizabeth, and Geraldine Pratt. "The Spectacular and the Mundane: Racialised State Violence and Filipino Migrant Families." *Environment and Planning A: Economy and Space* 44, no. 4 (2012): 889–904.

Levin, Laura, and Kim Solga. "Building Utopia: Performance and the Fantasy of Urban Renewal in Contemporary Toronto." *The Drama Review* 53, no. 3 (2009): 37–53.

Lo, Jacqueline, and Helen Gilbert. "Toward a Topography of Cross-Cultural Theatre Praxis." *The Drama Review* 46, no. 3 (2002): 31–53.

Lorente, Beatriz P. *Scripts of Servitude: Language, Labor Migration and Transnational Domestic Work*. Bristol: Multilingual Matters, 2017.

Lowe, Lisa. *The Intimacies of Four Continents*. Durham and London: Duke University Press, 2015.

Lupick, Travis. "CBSA Enforcement against Immigrants on the Rise in B.C." *Georgia Straight*, 13 January 2016. www.straight.com/news/614931/cbsa-enforcement-against-immigrants-rise-bc.

Macdonald, David. "Child Care Deserts." *Canadian Centre for Policy Alternatives*, 18 June 2018. www.policyalternatives.ca/publications/reports/child-care-deserts-canada.

Macklin, Audrey. "Foreign Domestic Worker: Surrogate Housewife or Mail Order Servant?" *McGill Law Journal* 37, no. 3 (1992): 681–760.

Manalanson IV, Martin F. *Global Divas: Filipino Gay Men in Diaspora*. Durham and London: Duke University Press, 2003.

Manalansan IV, Martin F. "Queering the Chain of Care Paradigm." *Scholar and Feminist Online* 6, no. 3 (2008): n.p. http://sfonline.barnard.edu/immgration/manalansan_01.htm.

Manalansan IV, Martin F., and Augusto F. Espiritu. "The Field: Dialogues, Visions, Tensions, and Aspirations." In *Filipino Studies: Palimpsests of Nation and Diaspora*, edited by Martin F. Manalanson IV and Augusto F. Espiritu, 1–11. New York: New York University Press, 2016.

Marcus, George. "The Ambitions of Theory Work in the Production of Contemporary Anthropological Research." In *Theory Can Be More Than It Used to Be*, edited by Dominic Boyer, James D. Faubion and George E. Marcus, 48–64. Ithaca, NY: Cornell University Press, 2015.

Marcus, George. "Art (and Anthropology) at the World Trade Organization: Chronicle of an Intervention." *Ethnos* 82, no. 5 (2017): 907–924.

Marcus, George. "The Legacies of *Writing Culture* and the Near Future of the Ethnographic Form: A Sketch." *Cultural Anthropology* 27, no. 3 (2012): 427–445.

Marschall, Anika. "What Can Theatre Do about the Refugee Crisis? Enacting Commitment and Navigating Complicity in Performative Interventions." *Research in Drama Education* 23, no. 2 (2018): 148–166.

Marston, Sallie, and Sarah de Leeuw. "Creativity and Geography: Towards a Politicized Intervention." *Geographical Review* 103, no. 2 (2013): iii–xxvi.

Martin, Carol. "Bodies of Evidence." *The Drama Review* 50, no. 3 (2006): 8–15.

Mawani, Renisa. *Colonial Proximities: Crossracial Encounters and Juridical Truths in British Columbia 1871–1921.* Vancouver: University of British Columbia Press, 2009.

Mbembe Achille. "Necropolitics." *Public Culture* 15, no. 1 (2003): 11–40.

McCallum, John. "CMI (A Certain Maritime Incident): Introduction." *Australasian Drama Studies* 48 (2006): 136–142.

McGrahahan, Carol. "Refusal and the Gift of Citizenship." *Cultural Anthropology* 31, no. 3 (2016): 334–341.

McGrahahan, Carol. "Theorizing Refusal: An Introduction." *Cultural Anthropology* 31, no. 3 (2016): 319–325.

McIvor, Charlotte. *Migration and Performance in Contemporary Ireland: Towards a New Interculturalism.* London: Palgrave Macmillan, 2016.

McKay, Deirdre. *Global Filipinos: Migrants' Lives in the Virtual Village.* Bloomington: Indiana University Press, 2012.

McKegney, Sam. "From Trickster Poetics to Transgressive Politics: Substantiating Survivance in Tomson Highway's *Kiss of the Fur Queen*." *Studies in American Indian Literatures* 17, no. 4 (2005): 79–113.

McKenzie, Jon. "Gender Trouble: (The) Butler Did It." In *The Ends of Performance*, edited by Peggy Phelan and Jill Lane, 217–235. New York: New York University Press, 1998.

McKittrick, Katherine. "Diachronic Loops/Deadweight Tonnage/Bad Made Measure." *Cultural Geographies* 23, no. 1 (2016): 3–18.

McKittrick, Katherine. "Mathematics Black Life." *The Black Scholar: Journal of Black Studies and Research* 44, no. 2 (2014): 16–28.

Meggs, Geoff. "Nanay: Filipino Word Meaning 'Mother.'" *Geoff Meggs: Vancouver City Councilor* (blog), 9 February 2009. www.geoffmeggs.ca/2009/02/09/nanay-filipino-word-meaning-mother/.

Mercene, Floro. *Manila Men in the New World: Filipino Migration to Mexico and the Americas from the Sixteenth Century.* Quezon: University of the Philippines Press, 2007.

Migrante International. "Community Profile." n.p. n.d.

Migrante International. "#SONA2015 Number of OFWs Leaving Daily Rose from 2,500 in 2009 to 6,092 in 2015." *Migrante International* (blog), 29 July 2015. http://migranteinternational.org/2015/07/29/sona2015-number-of-ofws-leaving-daily-rose-from-2500-in-2009-to-6092-in-2015/.

Mikdashi, Maya. "What is Settler Colonialism? (for Leo Delano Ames, Jr.)." *American Indian Culture and Research Journal* 37, no. 2 (2013): 23–34.

Miller, Stuart Creighton. *Benevolent Assimilation: The American Conquest of the Philippines, 1899–1903.* New Haven: Yale University Press, 1982.

Millner, Naomi. "Activist Pedagogies through Ranciere's Aesthetic Lens." In *Geographical Aesthetics Imagining Space, Staging Encounters*, edited by Harriet Hawkins and Elizabeth Straughan, 71–90. London: Routledge, 2015.

Mohanty, Chandra T. "Under Western Eyes: Feminist Scholarship and Colonial Discourses." *Boundary 2* 12, no. 3 (1986): 333–358.

Molinari, Nicole. "Bare Care: Working Within/Reworking Restructured Long-Term Care." Unpublished Master's Thesis, University of British Columbia, 2018.

Morgensen, Scott Lauria. "Unsettling Queer Politics: What Can Non-Natives Learn from Two-Spirit Organizing?" In *Queer Indigenous Studies: Critical Interventions in Theory, Politics, and Literature*, edited by Qwo-Li Driskoll, Chris Finley,

Brian Joseph Killey, and Scott Lauria Morgensen, 132–154. Tucson: University of Arizona Press, 2011.

Nagar, Richa, in journey with Parakh Theatre and Sangtin Kisaan Mazdoor Sangathan. *Hungry Translations: Relearning the World through Radical Vulnerability.* Champaign: University of Illinois Press, forthcoming.

Nagar, Richa. "Hungry Translations: The World through Radical Vulnerability." *Antipode* early view (2018): 1–22.

Nestruck, J. Kelly. "Dated Agitprop Muffles Caregivers' Battle Cry." *Globe and Mail*, 27 February 2010. www.theglobeandmail.com/arts/theatre-and-performance/dated-agitprop-muffles-the-battle-cry/article4187924/.

Ofreneo, Rene E. "Precarious Philippines: Expanding Informal Sector, 'Flexibilizing' Labor Market." *American Behavioral Scientist* 57, no. 4 (2013): 420–443.

Oke, Chris. "Yukon Retains Its Foreign Workers." *Yukon News*, 8 June 2011. www.yukon-news.com/news/yukon-retains-its-foreign-workers/.

Olsen, Cecile Sachs. "Collaborative Challenges: Negotiating the Complicities of Socially Engaged Art within an Era of Neoliberal Urbanism." *Environment and Planning D: Society and Space* 36, no. 2 (2018): 273–293.

Ong, Aihwa. *Neoliberalism as Exception: Mutations in Citizenship and Sovereignty.* Durham and London: Duke University Press, 2006.

Ong, Aiwha. "Women out of China: Traveling Tales and Traveling Theories in Post-Colonial Feminism." In *Women Writing Culture*, edited by Ruth Behar and Deborah Gordon, 350–372. Berkeley: University of California Press, 1995.

Organization for Economic Cooperation and Development, and the European Commission. "A Good Life in Old Age? Monitoring and Improving Quality of Long-Term Care." Policy Brief, Paris, June 2013. www.oecd.org/els/health-systems/PolicyBrief-Good-Life-in-Old-Age.pdf.

Organization for Economic Cooperation and Development. "OECD Family Database." Database, Paris, 2017. www.oecd.org/els/family/database.htm.

Paley, Dawn. "Olympics Cash and Vancouver's Cultural Community: Lines Are Being Drawn between Those Who Accepted Cultural Olympiad Money, and Those Who Refused It." *Vancouver Media Co-op*, 11 February 2010. http://vancouver.mediacoop.ca/story/2679.

Parreñas, Rhacel Salazar. *Children of Global Migration: Transnational Families and Gendered Woes.* Palo Alto: Stanford University Press, 2005.

Parreñas, Rhacel Salazar. "Long Distance Intimacy: Gender and Intergenerational Relations in Transnational Families." *Global Networks* 5, no. 4 (2005): 317–336.

Parreñas, Rhacel Salazar. "Transnational Mothering: A Source of Gender Conflicts in the Family." *North Carolina Law Review* 88 (2010): 1825–1856.

PATAMABA-IUP. "Surfacing HBW Issues and Local Level Responses as Entry Points for Organizing: Focus on Urban Communities in Two Cities of Metro Manila Philippines." Homenet South Asia, Mauritius, n.d. http://homenetsouthasia.net/pdf/Study_of_HBW's_issues_in_2_cities_of_Philippines.pdf.

Paul, Anju Mary. "Capital and Mobility in the Stepwise International Migrations of Filipino Migrant Domestic Workers." *Migration Studies* 3, no. 3 (2015): 438–459.

Paul, Anju Mary. *Multinational Maids: Stepwise Migration in a Global Labor Market.* Cambridge: Cambridge University Press, 2017.

Paul, Anju Mary. "Stepwise International Migration: A Multi-Stage Migration Pattern for the Aspiring Migrant." *American Journal of Sociology* 116, no. 6 (2011): 1842–1886.

Pearson, Mike. *Marking Time: Performance Archaeology and the City*. Exeter: University of Exeter Press, 2013.

Pedwell, Carolyn. *Affective Relations: The Transnational Politics of Empathy*. Basingstoke: Palgrave Macmillan, 2014.

Phelan, Peggy. *Unmarked: The Politics of Performance*. New York: Routledge, 1993.

Phelps, Jerome. "Why Is So Much Art about the 'Refugee Crisis' So Bad?" *OpenDemocracy 50.50*, 11 May 2017. www.opendemocracy.net/5050/jerome-phelps/refugee-crisis-art-weiwei.

Philippine Overseas Employment Administration. "Statistical Tables on Overseas Filipino Workers." Labor Force Statistics, Manila, 2018. https://psa.gov.ph/statistics/survey/labor-force/sof-index.

Pinches, Michael. "Modernisation and the Question for Modernity: Architectural Form, Squatters Settlements and the New Society in Manila." In *Cultural Identity and Urban Change in Southeast Asia: Interpretative Essays*, edited by Mark Askew and William S. Logan, 13–42. Geelong: Deakin University Press, 1994.

Pinder, David. "Arts of Urban Exploration." *Cultural Geographies* 12 (2005): 383–411.

Pinder, David. "Sound, Memory and Interruption: Ghosts of London's M11 Link Road." In *Cities Interrupted: Visual Culture and Urban Space*, edited by Shirley Jordan and Christopher Lindner, 65–83. London: Bloomsbury Academic, 2016.

Polanco Sorto, Aida Geraldina. "Behind the Counter: Migration, Labour Policy and Temporary Work in a Global Fast Food Chain." Unpublished PhD Dissertation, University of British Columbia, 2013.

Polanco Sorto, Aida Geraldina. "Consent Behind the Counter: Aspiring Citizens and Labour Control Under Precarious (Im)migration Schemes." *Third World Quarterly* 37, no. 8 (2016): 1332–1350.

Povinelli, Elizabeth. *Economies of Abandonment*. Durham: Duke University Press, 2011.

Pratt, Geraldine. *Families Apart: Migrant Mothers and the Conflicts of Labor and Love*. Minneapolis: University of Minnesota Press, 2012.

Pratt, Geraldine. *Working Feminism*. Philadelphia: Temple University Press, 2004.

Pratt, Geraldine, and Caleb Johnston. "Filipina Domestic Workers, Violent Insecurity, Testimonial Theatre and Transnational Ambivalence." *Area* 46, no. 4 (2014): 358–360.

Pratt, Geraldine, and Caleb Johnston. "Translating Research into Theatre: *Nanay*: A Testimonial Play." *BC Studies* 163 (2009): 123–132.

Pratt, Geraldine, and Caleb Johnston, in collaboration with the Philippine Women Centre of British Columbia. "Nanay (Mother): A Testimonial Play." In *Once More, With Feeling: Five Affecting Plays*, edited by Erin Hurley, 49–90. Toronto: University of Toronto Press, 2014.

Pratt, Geraldine, Caleb Johnston, and Vanessa Banta. "A Travelling Script: Labour Migration, Precarity and Performance." *The Drama Review* 61, no. 2 (2017): 48–70.

Pratt, Geraldine, Caleb Johnston, and Vanessa Banta. "Lifetimes of Disposability and Surplus Entrepreneurs in Bagong Barrio, Manila." *Antipode* 49, no. 1 (2017): 169–192.

Pratt, Geraldine with Migrante British Columbia. "Organising Domestic Workers in Vancouver Canada: Gendered Geographies and Community Mobilization." *Political Power and Social Theory* 35 (2018): 99–119.

Pratt, Mary Louise. "I, Rigoberta Menchú and the 'Culture Wars'." In *The Rigoberta Menchú Controversy*, edited by Arturo Arias, 29–48. Minnesota: University of Minnesota Press, 2001.

Puar, Jasbir, Lauren Berlant, Judith Butler, Bojana Cvejic, Isabell Lorey, and Ana Vujanovic. "Precarity Talk: A Virtual Roundtable with Lauren Berlant, Judith Butler, Bojana Cvejic, Isabell Lorey, Jasbir Puar, and Ana Vujanovic." *The Drama Review* 56, no. 4 (2012): 163–177.

Pulido, Laura. "Geographies of Race and Ethnicity III: Settler Colonialism and Non-Native People of Color." *Progress in Human Geography* 42, no. 2 (2018): 309–318.

Pulumbo-Liu, David. *Asian/American: Historical Crossings of a Racial Frontier*. Palo Alto: Stanford University Press, 1999.

Rafael, Vincente. "Your Grief Is Our Gossip: Overseas Filipinos and Other Spectral Presences." *Public Culture* 9 (1997): 267–291.

Rafael, Vincente. *White Love and Other Events in Filipino History*. Durham: Duke University Press, 2000.

Rancière, Jacques. *Disagreement: Politics and Philosophy*. Translated by Julie Rose. Minneapolis: University of Minnesota Press, 1999.

Rancière, Jacques. "Dix theses sur la politique." *Aux Bords du Politique*, 2nd edition. Paris: Editions Osiris, 1998.

Rancière, Jacques. *The Hatred of Democracy*. Translated by Steve Corcoran. London: Verso, 2006.

Rancière, Jacques. *The Politics of Aesthetics: The Distribution of the Sensible*. Translated by Gabriel Rockhill. London: Continuum, 2004.

Raynor, Ruth. "Speaking, Feeling, Mattering: Theatre as Method and Model for Practice-Based, Collaborative, Research." *Progress in Human Geography* early view (2018): n.p.

Razack, Sherene. *Dark Threats and White Knights: The Somalia Affair, Peacekeeping, and the New Imperialism*. Toronto: University of Toronto Press, 2004.

Richardson, Michael. "Theatre as Safe Space? Performing Intergenerational Narratives with Men of Irish Descent." *Social & Cultural Geography* 16, no. 6 (2015): 615–633.

Roach, Joseph. *Cities of the Dead: Circum-Atlantic Performance*. New York: Columbia University Press, 1996.

Robinson, Jennifer. "Comparative Urbanism: New Geographies and Cultures of Theorizing the Urban." *International Journal of Urban and Regional Research* 40, no. 1 (2016): 187–199.

Rodriguez, Dylan. *Suspended Apocalypse: White Supremacy, Genocide, and the Filipino Condition*. Minneapolis: University of Minnesota Press, 2010.

Rodriguez, Robyn Magalit. *Migrants for Export: How the Philippine State Brokers Labor to the World*. Minneapolis: University of Minnesota Press, 2010.

Rodriguez, Robyn Magalit. "Toward a Critical Filipino Approach to Philippine Migration." In *Filipino Studies: Palimpsests of Nation and Diaspora*, edited by Martin F. Manalanson IV and Augusto F. Espiritu, 33–55. New York: New York University Press, 2016.

Rogers, Amanda. "Advancing the Geographies of the Performing Arts: Intercultural Aesthetics, Migratory Mobility and Geopolitics." *Progress in Human Geography* 42, no. 4 (2017): 549–568.

Ronson, Jacqueline. "Film Fest Brings World Stories Home." *Yukon News*, 1 February 2013. www.yukon-news.com/sports/film-fest-brings-world-stories-home/.

Rosas, Gilberto. *Barrio Libre: Criminalizing States and Delinquent Refusals of the New Frontier*. Durham and London: Duke University Press, 2012.

Rowe, Aimee Carillo, and Eve Tuck. "Settler Colonialism and Cultural Studies: Ongoing Settlement, Cultural Production and Resistance." *Cultural Studies↔Critical Methodologies* 17, no. 1 (2017): 3–13.

Roy, Ananya. "What is Urban about Critical Urban Theory?" *Urban Geography* 37, no. 6 (2015): 810–823.

Ryan, Allan. *The Trickster Shift: Humour and Irony in Contemporary Native Art*. Vancouver: University of British Columbia Press, 1999.

Said, Edward. *Orientalism*. New York: Pantheon Books, 1978.

Saldaña, Johnny. *Ethnotheatre: Research from Page to Stage*. New York: Routledge, 2011.

Salverson, Julie. "Change on Whose Terms? Testimony and an Erotics of Inquiry." *Theater* 31, no. 3 (2001): 119–125.

Salverson, Julie. "Taking Liberties: A Theatre Class of Foolish Witnesses." *Research in Drama Education* 13, no. 2 (2008): 245–255.

San Juan Jr., E. *After Postcolonialism: Remapping Philippines-United States Confrontations*. Lanham: Rowman and Littlefield, 2000.

San Juan Jr., E. *Toward Filipino Self Determination: Beyond Transnational Globalization*. Albany: State University of New York Press, 2009.

Saranillio, Dean. "Why Asian Settler Colonialism Matters: A Thought Piece on Critiques, Debates, and Indigenous Difference." *Settler Colonial Studies* 3 (2013): 280–294.

Saunders, Doug. "Family Ties." *Globe and Mail*, 16 June 2018. O1, O6–O7.

Schecter, Tanya. *Race, Class, Women and the State: The Case of Domestic Labour in Canada*. Montreal: Black Rose Books, 1998.

Schiller, Nina Glick, and Ayse Çaglar. "Migrant Incorporation and City Scale: Towards a Theory of Locality in Migration Studies." *Journal of Ethnic and Migration Studies* 35, no. 2 (2009): 177–202.

Schneider, Rebecca. "New Materialism and Performance Studies." *The Drama Review* 59, no. 4 (2015): 7–17.

Sepulveda, Juan Manuel. *The Battle of Oppenheimer Park*. Mexico: Fragua Cine, 2016. Video recording.

Sharpe, Christina. *In the Wake: On Blackness and Being*. Durham and London: Duke University Press, 2016.

Shih, Shu-mei. "Comparison as Relation." In *Comparison: Theories, Approaches, Uses*, edited by Rita Felski and Susan S. Friedman, 79–98. Baltimore: The Johns Hopkins University Press, 2013.

Shih, Shu-mei. "Race and Relation: The Global Sixties in the South of the South." *Comparative Literature* 68, no. 2 (2016): 141–154.

Shih, Shu-mei. "World Studies and Relational Comparison." *PMLA* 130, no. 2 (2015): 430–438.

Silva, Denise Ferreira da. *Toward a Global Idea of Race*. Minneapolis and London: University of Minnesota Press, 2007.

Silversides, Ann. "Long-Term Care in Canada: Status Quo No Option." Canadian Federation of Nurses Unions, Ottawa, 3 February 2011. https://nursesunions.ca/wp-content/uploads/2017/07/long_term_care_paper.final__0.pdf.

Simpson, Audra. "Consent's Revenge." *Cultural Anthropology* 31, no. 3 (2016): 326–333.

Simpson, Audra. *Mohawk Interruptus: Political Life across the Borders of Settler States*. Durham and London: Duke University Press, 2014.

Simpson, Leanne B. *As We Have Always Done: Indigenous Freedom through Radical Resistance*. Minneapolis: University of Minnesota Press, 2017.

Singer, P. W. *Corporate Warriors: The Rise of the Privatized Military Industry*. Ithaca, NY: Cornell University Press, 2007.

Soans, Robin. "Robin Soans." In *Verbatim Verbatim: Contemporary Documentary Theatre*, edited by Will Hammond and Don Steward, n.p. London: Oberon Books, 2008.

Spivak, Gayatri Chakravorty. "Can the Subaltern Speak?" In *Marxism and the Interpretation of Culture*, edited by Cary Nelson and Lawrence Grossberg, 271–313. London: Macmillian, 1988.

Spivak, Gayatri Chakravorty. "Thinking Cultural Questions in 'Pure' Literary Terms." In *Without Guarantees: In Honour of Stuart Hall*, edited by Paul Gilroy, Lawrence Grossberg, and Angela McRobbie, 335–576. London and New York: Verso, 2000.

Stackhouse, John. "Comic Heroes or 'Red Niggers'?" *Globe and Mail*, 9 November 2001. http://v1.theglobeandmail.com/series/apartheid/stories/20011109-1.html.

Stanley, Anna, Sedef Arat-Koc, Laurie Bertram, and Hayden King. "Intervention – Addressing the Indigenous-Immigration 'Parallax Gap.'" *Antipode Foundation* (blog), 18 June 2014. https://antipodefoundation.org/2014/06/18/addressing-the-indigenous-immigration-parallax-gap/.

Statistics Canada. "2016 Census: City of Whitehorse, Yukon and Yukon Territory." Census Profile, Ottawa, 2017.

Stewart, Kathleen. *Ordinary Affects*. Durham and London: Duke University Press, 2007.

Swanson, Kerry. "The Noble Savage was a Drag Queen: Hybridity and Transformation in Kent Monkman's Performance and Visual Art Interventions." In *Queerly Canadian: An Introductory Reader in Sexuality Studies*, edited by Maureen Fitzgerald and Scott Rayter, 565–576. Toronto: Canadian Scholars Press, 2012.

Tabuga, Aubrey D. "How Do Filipino Families Use the OFW Remittances?" *Philippine Institute for Development Studies Policy Notes*, no. 2007–2012 (2007): 1–8.

Tadiar, Neferti X.M. "Decolonization, 'Race,' and Remaindered Life under Empire." *Qui Parle: Critical Humanities and Social Sciences* 23, no. 2 (2015): 135–160.

Tadiar, Neferti X.M. "Domestic Bodies of the Philippines." *Sojourn: Journal of Social Issues in Southeast Asia* 12, no. 2 (1997): 153–191.

Tadiar, Neferti X.M. "Life-Times of Disposability in Global Neoliberalism." *Social Text 115* 31, no. 2 (2013): 19–47.

Tadiar, Neferti X.M. *Things Fall Away: Historical Experience and the Making of Globalization*. Durham and London: Duke University Press, 2009.

Tagg, John. "Melancholy Realism: Walter Evans's Resistance to Meaning." *Narrative* 11, no. 1 (2003): 3–77.

TallBear, Kim. *Making Love and Relations beyond Settler Sexualities*. Uppsala: Technoscience Research Platform, 2016. Video recording. www.youtube.com/watch?v=rICFnEDRIts.

Taylor, Diana. *The Archive and the Repertoire: Performing Cultural Memory in the Americas*. Durham and London: Duke University Press, 2003.

Taylor, Diana. "Archiving the 'Thing': Teatro da Vertigem's Bom Retiro 958 Metros." *The Drama Review* 59, no. 2 (2015): 58–73.

Taylor, Diana. "Save As." *On the Subject of Archives* 9, no. 1 & 2 (2012): n.p. http://hemisphericinstitute.org/hemi/en/e-misferica-91/taylor.

Taylor, Drew Hayden. *Redskins, Tricksters and Puppy Stew.* Montreal: National Film Board of Canada, 2000. Video recording. www.nfb.ca/film/redskins_tricksters_puppy_stew/.

Thompson, John. "Territory to Expand Foreign Worker Program." *Yukon News*, 18 June 2010. www.yukon-news.com/news/territory-to-expand-foreign-worker-program/.

Thrush, Coll. *Indigenous London: Native Travelers at the Heart of Empire.* New Haven: Yale University Press, 2016.

Tiongson, Nicanor G. *Dulaan: An Essay on Philippine Theatre.* Manila: Cultural Center of the Philippines, 1989.

Todd, Douglas. "The Caregiver Conundrum: Caregiver Plan Popular, Problematic: Nine Simmering Debates about Canada's Approach to Foreign Domestic Workers." *Vancouver Sun*, 24 May 2014. A1, A6.

Tolentino, Rolando B. "Cityscape: The Capital Infrastructuring and Technologization of Manila." In *Cinema and the City: Film and Urban Societies in Global Context*, edited by Mark Shiel and Tony Fitzmaurice, 158–170. Oxford: Blackwell, 2001.

Tolia-Kelly, Divya P. "Rancière and the Re-distribution of the Sensible: The Artist Rosanna Raymond, Dissensus and Postcolonial Sensibilities within the Spaces of the Museum." *Progress in Human Geography* early view (2017): n.p.

Tuck, Eve, and K. Wayne Yang. "Decolonization Is Not a Metaphor." *Decolonization: Indigeneity, Education & Society* 1, no. 1 (2012): 1–40.

Tungohan, Ethel. *From the Politics of Everyday Resistance to the Politics from Below: Migrant Care Worker Activism in Canada.* Unpublished PhD Dissertation, University of Toronto, 2014.

Tungohan, Ethel. "Reconceptualizing Motherhood, Reconceptualizing Resistance." *International Feminist Journal of Politics* 15, no. 1 (2013): 39–57.

van Erven, Eugene. *Community Theatre: Global Perspectives.* London and New York: Routledge, 2001.

van Erven, Eugene. *Stages of People Power: The Philippines Educational Theatre Association.* Verhandelingen no. 43. The Hague: Centre for the Study of Education in Developing Countries (CESO), 1989.

Vizenor, Gerald. *Manifest Manners: Narratives on Postindian Survivance.* Lincoln: Nebraska, 1999.

Volpp, Leti. "The Indigenous as Alien." *UC Irvine Law Review* 5 (2015): 289–325.

Walcher, Werner. *Cold Paradise.* Yukon: Cold Paradise Productions, 2014. Video recording.

Wallis, Maria, and Wenona Giles. "Defining the Issues on Our Terms: Gender, Race and State – Interviews with Racial Minority Women." *Resources for Feminist Research* 17, no. 3 (1988): 43.

Wales, Prue. "Temporarily Yours: Foreign Domestic Workers in Singapore." In *Research-Based Theatre: An Artistic Methodology*, edited by George Belliveau and Graham Lea, 147–161. Bristol: Intellect, 2016.

Weiss, Erica. "Refusal as Act, Refusal as Abstention." *Cultural Anthropology* 31, no. 3 (2016): 351–358.

West Coast Domestic Workers' Association. "Temporary Foreign Worker Program: A Submission by the West Coast Domestic Workers' Association to the Standing

Committee on Human Resources, Skills and Social Development and the Status of Persons with Disabilities." 31 May 2016. http://migrantrights.ca/wp-content/uploads/2016/06/WCDWA-Brief-to-HUMA-Committee-for-TFWP-Review.pdf.

Wilmer, Stephen Elliot. *Performing Statelessness in Europe*. London: Palgrave Macmillan, 2018.

Wimmer, Andreas, and Nina Glick Schiller. "Methodological Nationalism and Beyond: Nation-State Building, Migration and the Social Sciences." *Global Networks* 2, no. 4 (2002): 301–334.

Wingrove, Josh. "Filipino Community Thrives in Yukon; Streamlined Path to Entry, Residency Appeals to Immigrants Who Come to Work." *Globe and Mail*, 23 January 2014. A4.

Wolfe, Patrick. "Settler Colonialism and the Elimination of the Native." *Journal of Genocide Research* 8, no. 4 (2006): 387–409.

Wong, Rita. "Decolonization: Reading Asian and First Nations Relations in Literature." *Canadian Literature* 199 (2008): 158–180.

World Bank Group. "Migration and Remittances: Recent Developments and Outlook." World Bank Migration and Development Brief 27, Washington DC, 2017. http://pubdocs.worldbank.org/en/992371492706371662/MigrationandDevelopmentBrief27.pdf.

Yang, Dean. "International Migration, Human Capital, and Entrepreneurship: Evidence from Philippine Migrants' Exchange Rate Shocks." World Bank Policy Research Working Paper Series No. 3578, Washington DC, 2005.

Index

For Product Safety Concerns and Information please contact our EU representative GPSR@taylorandfrancis.com
Taylor & Francis Verlag GmbH, Kaufingerstraße 24, 80331 München, Germany

www.ingramcontent.com/pod-product-compliance
Lightning Source LLC
Chambersburg PA
CBHW050450051126
54379CB00003B/85
9781138885639